Improving
the Quality of Life

IMPROVING THE QUALITY OF LIFE

A Holistic Scientific Strategy

Myles I. Friedman

Westport, Connecticut
London

Library of Congress Cataloging-in-Publication Data

Friedman, Myles I., 1924–
    Improving the quality of life : a holistic scientific strategy /
Myles I. Friedman.
    p. cm.
    Includes bibliographical references and index.
    ISBN 0–275–96028–5 (alk. paper)
    1. Quality of life. 2. Social indicators. 3. Social psychology.
I. Title.
HN25.F75   1997
306—dc21   97–8839

British Library Cataloguing in Publication Data is available.

Copyright © 1997 by Myles I. Friedman

All rights reserved. No portion of this book may be
reproduced, by any process or technique, without the
express written consent of the publisher.

Library of Congress Catalog Card Number: 97–8839
ISBN: 0–275–96028–5

First published in 1997

Praeger Publishers, 88 Post Road West, Westport, CT 06881
An imprint of Greenwood Publishing Group, Inc.

Printed in the United States of America

The paper used in this book complies with the
Permanent Paper Standard issued by the National
Information Standards Organization (Z39.48–1984).

10 9 8 7 6 5 4 3 2 1

Copyright Acknowledgments

The author and publisher are grateful for permission to reproduce the
following copyrighted material:

Table 2.18, "Life Situation Survey," from Robert A. Chubon, Ph.D.,
University of South Carolina. Copyright © 1984 by Robert A. Chubon.

Table 2.19, "Names of Quality of Life Instruments Used in the 75
Articles Reviewed," from Gill, T.M., and Feinstein, A.R. (1994). A
critical appraisal of quality-of-life measurements. *Journal of the
American Medical Association* 272: 622. Copyright © 1994, American
Medical Association.

To Betty

Improving the quality of life may be
the only goal people unanimously endorse
and willingly contribute to achieving.

Contents

Preface	xi
Acknowledgments	xiii
I. Fundamentals	1
1. Basic Issues and Challenges	3
2. Data to Build Upon	19
II. A Strategy for Improving the Quality of Life	59
3. Quality of Life as a Field of Study	61
4. Pursuing Improvements Scientifically	91
5. Working with Human Motivation	117
6. Developing Intelligence: The Means to Many Ends	137
7. The Holistic Approach	157
References	179
Index	189

Preface

Improving the quality of life is, of course, the focus of the book, as the title indicates. The emphasis is on improving the quality of life, not just advancing knowledge. People have a growing interest in improving the quality of their lives, and professionals are more dedicated than ever to helping them, as will be shown in Chapters 1 and 2. There is a pressing need to build on the disjointed and diverse research data on quality of life so that professionals might be more effective in their efforts to improve the quality of people's lives. It takes more than knowledge to help people—it takes the application of know-how.

A holistic approach to improving the quality of life is prescribed, as the subtitle indicates, for several reasons:

1. It has become clear that many different factors can contribute to quality of life.
2. A number of professionals may need to work together to improve the quality of a person's life.
3. People function as whole human beings. If they are to be fully understood and helped, they need to be observed and treated holistically.
4. There is a need to identify and circumscribe quality of life as a body of knowledge and a field of study.

The advantages of a holistic approach have been recognized by many professionals. However, it seems that most do not have the time or orientation to do much about it. They are taxed keeping up with advancements in their specializations and tend to view quality of life from the biased vantage point of their specialization. One aim of the holistic approach is to overcome the professional myopia that subverts efforts to help people. Although many professionals have advocated a holistic approach to quality of life, so far there has been more talk than effort. Nothing that qualifies as a holistic paradigm could be found in publication.

A scientific strategy for improving the quality of life is also proposed, as the subtitle implies. The traditional scientific method has been considered by many scientists to be sacrosanct, representing the essence of science. It has been taught throughout the world in much the same way over the years, although each teacher may embellish it in his or her own way. However, the traditional scientific method has its limitations, and it is quite possible to conduct scientific research without adhering strictly to the scientific method. One need only honor the basic tenets of science. Many scientists working on research and development projects adapt the scientific method to their needs without compromising the scientific validity of their work.

The proposed scientific strategy abides by the mandates of science while deviating from the traditional scientific method. The strategy is an adaptation of scientific inquiry designed specifically to improve the quality of life. There are, of course, other approaches to quality of life that are not being pursued in this book, such as theological, philosophical, and humanistic approaches. Despite its limitations, science has in the past contributed to raising the standard of living. It can be applied constructively to improve the quality of life.

There is another unique feature of the proposed strategy. It includes prescriptions for dealing with psychological factors that affect quality of life. Efforts to improve quality of life often fail because potent psychological forces are overlooked or ignored. Researchers and clinicians quite often cannot, or do not, take into account psychological factors that determine the behavior of the people with whom they are working. Scientific inquiry adheres to laws of logic. Human behavior, to a great extent, is determined by psychological laws. Professionals can be more effective in helping people if they learn to take into account psychological forces that determine human behavior and achievement, primarily human motivation and intelligence.

No other work was found that presents a holistic approach, a scientific strategy, and prescriptions for working with psychological forces to improve the quality of life.

Acknowledgments

I would like to acknowledge the assistance given to me by Lorin Anderson, Karen Clayton, Christine McCormick, Richard Hohn, John Johnson, Robert Steward, and Valerie West. Their comments and suggestions not only helped me to correct inaccuracies; they enabled me to reorganize the text to highlight ideas. I would like to thank James T. Sabin, for his guidance in making the presentation more focused, parsimonious, and cogent, and Robert Chubon, for acquainting me with literature on quality of life and offering invaluable suggestions. Most of all, I want to thank my wife Betty for helping me write the book day by day and giving me her enduring devotion and support.

Part I

Fundamentals

To fully understand the holistic scientific strategy for improving the quality of human life presented in Part II, it is necessary to understand the fundamental issues and challenges that gave rise to the quality of life movement and the development of the strategy. These issues and challenges are highlighted succinctly here and clarified in Chapter 1.

1. *Public versus professional interests in quality of life.* Although both the public and professionals are interested in the quality of life movement, their interests are not the same.
2. *Specialized versus cooperative treatment of people.* Although professional specialization is necessary, cooperative treatment is often required to serve people's best interests.
3. *The instinctive drive for self-preservation versus the pursuit of life enhancement.* Although people are instinctively driven to preserve themselves, they are also interested in working to make their lives better in the future than they are in the present.
4. *Quantity versus quality of life.* Although people want to extend their lives, they have second thoughts when the quality of their lives is seriously compromised.
5. *Impulsive versus rational decision making.* Although there are subliminal desires prompting impulsive decision making, improving the quality of one's life most often requires reflective rational decision making.
6. *Parts versus whole approach to people's problems.* Although an ailment such as appendicitis requires that a particular body part be attended to, many solutions to human problems require a holistic approach.
7. *Traditional versus modified scientific method.* Although the traditional scientific method has been responsible for advancements in knowledge, it needs to be adapted to solve quality of life problems more effectively.

8. *Wholesome versus perverse attempts to improve quality of life.* Although elevating the quality of life can be a noble mission, it can and has run amok.

These eight issues and others need to be addressed and understood if we are to be successful in improving the quality of life. The proposed strategy presented in Part II undertakes to resolve these issues. It merits your consideration.

To begin the development of a new strategy, it is beneficial to profit from, and build on, previous research and development on quality of life. Chapter 2 provides the baseline data necessary for the development of the strategy. A survey of available research on quality of life indicators suggests basic ingredients of quality of life. Although fragmentary, the data provide the basis for circumscribing and identifying quality of life as a field of study in Chapter 3. Some indicators are commonly identified; others are not encountered frequently but may be equally important. It is challenging to consider the various indicators that have been identified to date, the variance among them, and the controversies about them.

Chapter 1

Basic Issues and Challenges

The purpose of this book is to present a holistic scientific strategy, based on prior research, for improving the quality of life so that professionals can be more effective in their efforts to help people. There is both the need and the demand for more effective ways to elevate the quality of life.

In recent years, improving the quality of life has surged to the forefront of human interest. People are more interested in quality of life issues now partly because increased longevity and the invention of more sophisticated lifesaving techniques force them to deal with quality of life issues more frequently. More people are living with disabilities; more people are deciding whether or not to use life-support systems to sustain the life of a severely incapacitated loved one; and more people are leaving living wills and health care powers of attorney.

There is another, more salutary reason for the growing interest in quality of life: More is known about ways to improve the quality of life than ever before. And people are eager to learn about them and to take advantage of them as they become available. More people are joining fitness clubs, reading health publications, retreating to spas, taking courses, and joining discussion and support groups to elevate the quality of their lives. Furthermore, their interests in quality of life are not always self-centered.

Most people spend a great deal of time working to improve the quality of other people's lives in some way, whether it be raising children, on the job, or promoting causes such as education or eradicating a disease. Allegedly, Freud said that people are both better and worse than they think they are. War, crime, and destruction are expressions of the worst side of human nature. Working to improve the quality of people's lives is an expression of the better side. It may be that the human race is at its very best when people work together to improve the quality of life. And improving the quality of life may well be the common objective on which people with diverse interests can cooperate willingly to achieve.

In addition to the public, the government and professionals pledged to help people are now placing increased emphasis on improving the quality of life, as illustrated by the following policy statement:

The National Institutes of Health (NIH) funds research and development to improve the nation's capability to prevent disease, improve health, and enhance the quality of life. Because so many life-threatening diseases have been eliminated or made preventable, the afflictions of non life-threatening chronic conditions are absorbing an increasing share of health care costs and research attention. Prevention and treatment of such conditions aim at improving the quality of life as well as extending the length of life. (National Institutes of Health 1990)

Dr. Murray Freed, past president of the American Academy of Physical Medicine and Rehabilitation, clarifies the issues:

The past four decades have witnessed the development of major artificial life-support systems: mechanical ventilators, renal dialysis and alimentation. Evacuation and emergency medical care have also played a significant role. Thus medicine and surgery have prevented the deaths of countless persons who were critically ill. Many have been cured. But a progressively increasing number have been left with chronic disabilities, and a high proportion have not been restored to an optimal level of function. A multiple-sample survey has revealed that only between 5% and 10% of the total population falls into this category of disability. This small number has resulted from increasing attention given to the process of improving the functional capacity of humankind in the event of handicap or disability. . . .

We all agree that the competent practice in our specialty should strive to improve the quality of life of patients. Though there is little question about the dedication of medicine to improving the quality of life, the questions remain: Of what does quality of life consist? Who determines quality of life? What are the effects of such judgments on the care provided the patient? (Freed 1984)

To begin to answer these questions, it is necessary to understand fundamental outcomes people strive to achieve.

BASIC PURSUITS

To understand the nature of basic human pursuits, two important driving forces must be considered: self-preservation and enhancement. People are driven to preserve their lives, and having preserved them, they are driven to enhance their lives, to make the future better than the present.

Self-Preservation

Self-preservation is often studied from a Darwinian perspective, where the survival of the fittest is investigated. It is often studied as a legal right, when issues such as self-defense are considered. But self-preservation is seldom studied as a human desire that has momentous implications for the quality of life. This neglect needs to be redressed. Many of the manifestations of self-preservation to be discussed will be familiar. However, they will be viewed and analyzed in a different context—as meaningful expressions of the desire for self-preservation, and in this context, they provide new insights into human nature.

The desire for self-preservation is expressed in many overt and arcane ways. The desire to stay young is openly, often flagrantly expressed. Vast sums of money are spent on rejuvenation treatments, ranging from hair coloring to cosmetic surgery of various kinds.

Literary classics tell fascinating stories about people who go to great lengths to stay young. *The Picture of Dorian Gray* expounds on the theme from a male perspective, and *Sunset Boulevard*, from a female perspective. It seems that people can be swayed by primitive instinctive wishes to believe that they can avert aging and live a vital and everlasting life.

The drive for self-preservation also makes people quite vulnerable to life-extension promotions that pander to people's susceptibility. Vitamins, minerals, bee pollen, organic foods, ginseng tea, meditation, self-hypnosis, yoga, and exercise are advocated to prolong life. Moreover, these treatments are not as easily debunked as youth preservation treatments. Treatments like cosmetic surgery and cosmetic makeup are by their very nature superficial, and people can see through "image is everything" promotions. In contrast, life-extension treatments can be given an aura of credibility. The convincing but tenuous argument often proffered is that the treatments extend life by improving health, when improvement of health by no means ensures life extension.

Life-extension promotions include more than health improvement treatments. They include the promotion of doctrines that explain how life extends into the hereafter and how to conceive of and prepare for life after death. Some doctrines explain how a person passes to another place in the hereafter, such as heaven or hell, with an interim stop in purgatory. These doctrines often prescribe how to earn passage to heaven, paradise, or nirvana. Other doctrines explain how humans transmigrate to another form in the hereafter rather than to another place. They are reincarnated as another living thing. Some people have recollections of many previous existences, and their accounts of adventures in past lives can be fascinating. Formal doctrines and organizations explain reincarnation as a part of the scheme of things and how to fit into it.

It seems that embedded in the drive for self-preservation is an inherent, deep-seated wish for and sense of our own immortality that is manifested in uncanny ways. For example, it is impossible for people to conceptualize their own terminal demise. When they try, they perpetuate their own existence as an observer witnessing a funeral; both they as observers and the funeral continue to exist. People don't seem to be able to conceive of themselves as not existing. However, they can and do conceive of their own continuing existence in very elaborate and imaginative ways.

Procreation is another way people seem to express and satisfy their desire for continuing existence. Although in procreation people are not continuing their own lives after death, they are in a sense living on in their progeny. They live on in at least two ways: They perpetuate their genetic traits, and as teachers, their values, skills, and knowledge are passed from generation to generation.

There is another way people can and do contribute to future generations to satisfy their desire for immortality. Their social contributions can be preserved and passed on from generation to generation. Humans, unlike lower creatures, record the creative works of their contemporaries and store them in galleries, museums, libraries, and data banks to benefit present and coming generations. People in a very real sense achieve immortality when their contributions are preserved for posterity. The contributions of Freud, Einstein, Mozart, Rembrandt, Pasteur, and many other creative geniuses live on and on. However, one does not have to be a creative genius to have his or her ideas published. People contribute to posterity when they publish a book or an article that is worthy of being stored in public libraries.

Producing children in an overcrowded world is not considered to be a social contribution. Producing beautiful artwork, discovering new phenomena, and inventing new products and procedures that improve the human condition are cherished social contributions. These sublimated creative expressions are deeply appreciated and immortalized.

In one way or another, knowingly or unwittingly, people seem to seek fame to perpetuate themselves. They do what they can to be recognized, honored, rewarded, and seen in the public eye. They enter contests, attach themselves to luminaries, and attempt to show their art in galleries and to have their music, prose, and poetry published. The rich but untalented try to immortalize themselves by having something named after them. Universities name libraries, stadiums, scholarships, fellowships, buildings, rooms, and chaired professorships for donors who sponsor them. And, too, people seek public office to be recognized and then higher offices to become more prominent. Although becoming famous may be one way of improving life, it is also a sure way of living on in the annals of history.

Pursuing longevity, rejuvenation, life in the hereafter, and everlasting fame can be beneficial or detrimental to life on earth in the here and now. Living in accordance with one's beliefs enhances mental health. It gives meaning and purpose to life, which in turn improves outlook and increases zest for life. Various doctrines prescribe the "good life" on earth and often the pathway to the good life after death. Living according to the prescriptions ascribes value to one's lifestyle and is uplifting. There is evidence indicating that people feel more fulfilled when they live in accordance with their beliefs. There is little, if any, evidence indicating that it is more fulfilling to adopt one set of beliefs rather than another.

On the other hand, attempts to stay young and extend life can be harmful. There are spiritual doctrines that advocate dangerous practices. For instance, when medical cures are available, prolonged attempts at spiritual healing can endanger life. And some doctrines incite transgression and belligerence against nonbelievers. In addition, radical diets can deprive people of essential nutrients, and excessive strenuous exercise can cause heart attacks, sprains, fractures, and other ailments. And cosmetic surgery can be dangerous. Breast implants have threatened women's lives, and some cosmetic facial surgery has left patients with disastrous outcomes.

Enhancement

People not only strive to preserve their lives; they want to enhance their lives as well. Enhancement extends beyond preservation. The desire for enhancement is manifested in people's aspirations to make their lives and the lives of their loved ones better in the future than they are in the present. The ways in which people attempt to enhance their lives are almost limitless, depending on their needs, ambitions, and wishes. In free countries, people are encouraged to pursue their aspirations, whatever they may be, and many examples have been cited of people who rose from humble means to the top. Aspiration is considered to be both the motivating force and the beginning of the road to success.

Although people may aspire to anything, there are some common, often mundane, aspirations that people share. People aspire to be healthy beyond self-preservation. They work to improve their strength, endurance, and agility, in general, and to become better in the performance of athletic and sports feats. They also strive to improve their mental health. They engage in yoga, meditation, support and counseling groups, and spiritual pursuits to enhance their outlook on life.

People are also interested in enhancing their work life by improving their earnings, work satisfaction, and status. They often make career changes for any or all of these reasons.

Many aspire to further their education and provide for the education of their children. Some want more education for its own sake. They want to be more knowledgeable, learned, and wise. Others see education as the means to other ends. Additional education is often needed to fulfill work ambitions. And education is a help to social climbers, should one wish to keep up with or surpass the Joneses.

In addition, many people are preoccupied with increasing their possessions. They may want more and finer clothes and cars. They may want a larger home in a more affluent neighborhood, their own airplane or boat, or their own stable of racehorses.

Many people seem to have strong recreational interests, stronger than their work interests. Some can't wait to retire so that they will have more time to fish, play bridge, spoil their grandchildren, or indulge themselves in some other way. Others may wish to travel to remote places that were not accessible to them when they had less free time and money.

Finally, people who are deprived of pursuing their aspirations want more freedom. People want to know their limits under the law and the prevailing restrictions imposed on them by the social or work groups they are a part of. But they want the freedom to be themselves and to pursue their aspirations. Moreover, whatever amount of freedom they may have, they will resent any reductions in their freedom of choice and feel oppressed. The desire for freedom makes people appreciate governments and laws that provide for freedom of choice and expression. And people will quite naturally want to emigrate from places that oppress them to places that provide more freedom.

Although the desire for self-preservation may most often take precedence over the desire for enhancement, this is not always the case. People may be willing to make great personal sacrifices, even die, for a cause they believe in. It is not unusual for people to die for their country or a spiritual doctrine. It is ironic that people's desire for self-preservation spawns the desire to pursue life everlasting in accordance with a spiritual doctrine they believe in, a doctrine for which they may be willing to sacrifice their lives to honor and preserve.

It is crucial for professionals working to improve the quality of people's lives to understand that people want to enhance as well as preserve their lives. For far too long, professionals in fields such as medicine have concentrated on curing people without fully realizing that they were focusing on preserving life while often neglecting to enhance it. They have begun to realize, however, that both preservation and enhancement must be promoted to improve the quality of life and are making greater efforts to learn about enhancement and how to treat people to better their lives.

Quality of Life Issues

The desires for self-preservation and enhancement are both potent driving forces that affect quality of life. However, in advanced modern societies, survival is for most people not too difficult to achieve. Other forms of self-preservation, such as perpetuating life here on earth indefinitely, are impossible to achieve. And it is presently impossible here on earth to determine objectively whether life after death can be achieved.

As indicated, the instinctive desire for self-preservation can cause people to make irrational decisions that can lead to tragic consequences. Elevating the quality of life on earth, whether self-preservation or enhancement is the goal, frequently requires curbing instinctive drives in favor of reflective, deliberate decision making based on learning. It appears that adult humans spend a great deal of time reconciling conflicts between primitive urges for self-preservation and their rational sense of what they must do to improve the quality of their lives.

One of the advantages of being human is being able to profit from the knowledge passed on from previous generations, know-how that enables humans to dominate the planet. Although human infants are among the most helpless of creatures, and although it takes them longer to mature than lower animals, by adulthood they have learned from parents and other teachers how to accomplish more by far than the most intelligent ape. At birth the behavior of all creatures is directed primarily by the instincts they inherit. The behavior of lower creatures continues to be dominated by their instincts throughout their lives. In contrast, by the time humans reach adulthood, the preponderance of their adaptive behavior is learned. And the behavior they learn is in large measure responsible for their superiority over other creatures and their success in society.

Another advantage of being human is that instead of following the dictates of their instincts impetuously, people can behave purposefully by using their superior intelligence to predict the consequences of their actions before they act and then choosing to pursue the outcomes they prefer to achieve. And they often prefer to pursue other than instinctive gratification. Rather than giving in to their inherent desire to eat when they are hungry, many people fast to lose weight and to honor religious sacraments. Rather than giving vent to their innate sexual impulses when they are sexually aroused, many people choose to abstain from sex to avoid venereal disease or to satisfy sacred religious or matrimonial vows.

The behavior people share in common with lower creatures is instinctive behavior. The behavior that elevates them above other animals, behavior that is especially human, is purposeful behavior based on learning. As wisdom develops with rationality and learning, people realize that they are not slaves to their innate drives. They can control their impulses and purposefully pursue their enlightened self-interests. They come to understand that there are no fountains of youth, that wishing won't make it so, that what they wish for may not be good for them, and that the consequences of wish fulfillment may not be what they had hoped for.

As we know, the decision to use or not to use life-support devices is frequently made under the most trying conditions. A patient may be in great pain or in a chronic vegetative state with no prospect of leading a normal life or, in some cases, no hope of regaining consciousness. The more people learn about or are faced with these decisions, the less convinced they become that extending life unconditionally is what they want. Although people have an innate lust for life, after informed rational deliberation, they are coming to the conclusion that *quality* of life is more precious than *quantity* of life. The question that arises when people think about the conditions under which human life should be prolonged is: When is the quality of life so compromised that life is not worth living?

Quality of life is always a major concern when a person's functional abilities are seriously impaired and cannot be fully restored; and this happens much more frequently now because of advancements in emergency medical care (Gorovitz 1982). As a result, severely disabled people survive who previously would have died. But many survive with profound disabilities and cannot be completely rehabilitated. Since these patients cannot be restored to their former selves, doctors who have been traditionally trained to "cure" people have had to revise their approaches. After discussing the patients' handicapping conditions with them, doctors nowadays ask patients what they hope to achieve, given their limitations, and a treatment plan is formulated taking the patients' aspirations into account (Rusk 1964, 1972). Handicapped people who were forsaken and permanently institutionalized a short while ago are now leading full and productive lives in the mainstream of society because of modern rehabilitation procedures and increased accommodation and opportunity provided for handicapped people.

Many among the elderly also have permanent functional disabilities. Their mobility, strength, endurance, or coordination might be impaired, substantially diminishing the quality of their lives. Treatment plans to help the elderly are also complex, extending far beyond medication. Recreation, occupational therapy, exercise, special diets, and group discussions are often part of their treatment. And to keep them vital and to prevent them from becoming despondent, they are encouraged and helped to retain as much control over their lives as possible.

The need to provide extended care for disabled people quite naturally expands quality of life concerns. When considering placing loved ones in a residential facility to provide the support they need, quality of life in the facility becomes a very personal and important concern. Hopefully, the support services offered in the facility will improve the quality of their lives. But many concerns come to mind. How qualified is the staff? Is there ample staff to care for the clientele? What provision is made for medical care? Are there nurses and doctors in residence or on call? What is the clientele like? What kind of recreation is provided? Are troublesome patients overly restrained or sedated? How regimented is the clientele? How much control of their own affairs is the clientele allowed? Is the clientele allowed privacy (Corden 1990)?

It is quite difficult to assess the appropriateness of a residential facility for a particular person. Some important information seems to be asked routinely. Other relevant information is often overlooked. A case in point concerns an elderly grandmother who was physically frail but mentally alert. Although the family did thorough evaluations of facilities and staff before choosing a nursing home for her, she was most unhappy with the one they chose. Most of the clientele suffered from mentally debilitating illness, so she was unable to find the companionship she needed. Moreover, she was unable to attend a church of the religious denomination that had been such an important part of her life for so many years.

Of course, the family may try to care for a disabled family member at home, in which case the quality of life of the family becomes an issue. In many cases, care at home by family tends to improve the quality of life of a disabled person. But gains in the quality of life of a disabled family member quite often result in reductions in the quality of life of caregiving family members (Parker 1990). Caregiving is a burdensome responsibility and reduces opportunities for other things, for example, reading, watching television, going out socially, and entertaining at home (Baldwin and Gerard 1990). In addition, social relations within the family can be affected. A parent may need to neglect other family members to care for a disabled child or spouse. Quite often, caring for a disabled person can be a financial strain to a family. Most often, raising children is a joy as well as a burden. Caring for a disabled loved one at home is a responsibility and a duty with few redeeming compensations.

Discussing disability draws attention to the dreary side of life. It causes us to think about factors that diminish quality of life. On the other hand, there is a bright side to the quality of life issue. In America and other free and abundant societies, people have a great many opportunities to improve the quality of life and the liberty to take advantage of the opportunities. There are, of course, pockets of poverty in rural areas and inner cities, but even the poorest people have the opportunity to better themselves.

Furthermore, state and federal governments subsidize housing and provide food stamps and health care to help the poor maintain a minimal quality of life. Past presidents of the United States have made heroic efforts to elevate quality of life. Franklin Delano Roosevelt conceived the New Deal to bring the country out of the Great Depression. Lyndon Johnson had visions of raising life in America to new heights when he conceived the Great Society, bringing us Head Start and Community Action Programs. And the United States and the United Nations provide funds and services to improve the quality of life in underdeveloped countries (Szalai and Andrews 1980).

Anyone who has access to world news learns about the advantages of life in Switzerland, the United States, the United Kingdom, and other modern societies. People can learn about the latest in transportation, communications, health care, education, high-tech employment, and consumer products and know that they are available. They can also learn of the advantages of living in a free society, freedom of choice, presumed innocence until proven guilty, and guaranteed legal representation. These can give hope to even the most oppressed, poverty-stricken people. Their innate nature causes them to wish for and dream about ideal states such as heaven, nirvana, paradise, Shangri-la, the promised land, happiness. On the other hand, they learn from their daily experiences and deliberations what they can do to improve the quality of their lives on earth, and when they can see their way clear, they usually pursue it. They try to improve their lives where they are, and when there are superior opportunities elsewhere, they migrate toward them.

Although people may hope to reach some ideal utopian state in this life or in the hereafter, most are acutely aware of the differences between their dreams and their daily lives. They can tell when they are awake, and they can distinguish their night and day dreams from their daily routines. People who can't are probably insane. Furthermore, people who achieve success in this world don't let their dreams interfere with their daily functioning and obligations. They take care of business.

Quality of life is a mundane concept that is on people's minds daily. They think about it and talk about it. Anywhere in the world, people who haven't seen each other for a while inquire about the well-being of the other. In addition, people are frequently concerned with deficits that can affect the quality of life. When they are ill, they want to get well. When they are short of money, they try to increase their income and may borrow money to tide them over. When their eyesight begins to fail, they buy glasses. When people get fat, they are concerned with losing weight.

One way or another, quality of life is always a concern. When people try to determine whether their lives or those of others they may be discussing are worth living, they are bound to discuss quality of life. Or when they think about ways to improve their lives in the future, they contemplate the quality of their lives. As one might expect, it was the public's growing interest in quality of life beginning in the 1960s (Michener 1970) that led to recent professional movements to improve quality of life.

Although people appreciate the right to pursue happiness, many have difficulty coping with such concepts. They find them too amorphous, abstract, and vague to work with pragmatically and willingly leave them for jurists, philosophers, and theorists to grapple with. In their daily lives, people are much more concerned with improving the quality of their lives in particular, concrete ways. And as they strive for improvement, they have a sense of the deprivations and sacrifices they are willing to tolerate.

Still, efforts to deal with quality of life issues have not always been humane and constructive. In post–World War I the German euthanasia program was conceived to improve the quality of life in Germany. Beset by the hardships of defeat and poverty, prominent lawyers, physicians, politicians, economists, and others began advocating eliminating people who were distressed, beyond help, and devoid of value as an act of mercy and moral responsibility. This rationale did not originate with the Nazis, and it is not unlike the reasoning of Kevorkian and other present-day advocates of mercy killing. The program began informally with the destruction of malformed infants. It was extended to adults with physical disabilities and mental retardation. The gas chambers and crematoriums built in mental institutions to dispose of a large number of people said to be incurably insane were precursors of Auschwitz and other extermination centers of the Holocaust. Those of us working to improve the quality of life need to keep in mind that the noblest of intentions can run amok. To learn more about the German euthanasia programs, see Binding and Hoche (1975), Wolfensberger (1981), and Proctor (1988).

Approaching Quality of Life Holistically

Quality of life is a word label for a category. Like all categories, it is a holistic designation defined by the attributes of categorical membership. For instance, *mammal* is the word label of a category whose defining attributes are (1) warm-blooded, (2) hairy, (3) vertebrates (4) whose females suckle their young. The basic challenge is to specify the attributes of quality of life so that people and professionals will know what to do to improve the quality of life. Attributes that are presently measured to assess quality of life are identified in Chapter 2, where research on quality of life indicators is reviewed. The disparate attributes of quality of life revealed by the review are integrated in Chapter 3 to define quality of life as a field of study.

Specifying the attributes of quality of life is necessary but insufficient for improving the quality of life. To improve quality of life, desired improvements need to be inferred from the derived list of attributes. Attributes merely identify ingredients of quality of life. Improvements specify the direction of movement and goals that need to be achieved. Sometimes improvements can be inferred from attributes easily. For instance, for the attribute "functional ability," it is evident that greater functional ability is the desired improvement. On the other hand, on the pleasure-pain continuum, it is evident that pain reduction is most often a desired improvement—but not always. Pain indicates problems, and it is important to know when one is in trouble. Moreover, a state of extreme pleasure or euphoria is not often desirable. Drug-induced euphoria and manic states, in general, can be detrimental.

A holistic approach to quality of life extends even further. Treatments must be identified that can bring about desired improvements. And treatment plans are often multifaceted. Surgery, medication, vocational counseling, physical therapy, and job training may need to be combined to treat a disabled person, and special equipment may need to be provided to raise the quality of a person's life to an acceptable level. A motorized wheelchair and a modified van may be needed. In addition, a number of professionals may need to cooperate in administering different treatments.

Furthermore, people view themselves holistically. They have a holistic sense of self that consists of more than a summation of their various parts, acts, desires, and beliefs. They function as a whole person in their work, social, and recreational lives. And they are often piqued when others overemphasize a particular facet of their being, a particular talent or fault. They resent being disintegrated. They want to be appreciated for themselves as a whole, despite their particular strengths and weaknesses.

People also resent being compartmentalized and categorized by highly specialized clinicians. They want to be treated humanely, with dignity, as a whole person. They expect their doctors, dentists, accountants, and other professionals to expand their concerns and interests. They want to be told how a prescribed treatment will affect the quality of their lives—their family, work, recreational, and sex lives. They want to know more about the side effects of medication. Will it make them nervous, drowsy, impotent, depressed?

A holistic approach is sorely needed to overcome the professional tunnel vision caused by proliferating specialization, as evidenced by the increased number of classifications in the yellow pages of phone directories over the past 70 years. Moreover, the classification of professionals, such as doctors, keeps splintering.

Professional isolation seems to be partly responsible for the restricted purview of specialists. Professionals all too seldom work with professionals in other fields. Many exchange ideas primarily with professionals in their own fields, read the same publications, and attend the same meetings. Their identities and self-worths are deeply rooted in their specializations, and they feel uncomfortable straying too far from their fields of expertise.

While the complexity of human problems makes specialization important, it is equally important that professionals know enough about other specializations to make needed referrals and to work as team members when appropriate treatment requires the cooperation of a number of professionals in different fields. The need for different professionals to work together to treat people adequately and the need for each professional to know how a particular treatment affects the functioning of the whole person have forced specialists to expand their perspectives. Furthermore, when people are severely disabled, with no hope of recovery, concern shifts from treating particular symptoms to the overall quality of their lives.

Specialists that focus on one factor in isolation lose perspective and context and therefore must suffer in their efforts to understand and help people. Professionals who try to deal with overall quality of life without considering the factors that contribute to it are being superficial and also must suffer in their efforts to understand and help people.

Many professionals treating people not only need to be concerned with the whole person to the extent possible; they need to be concerned with the person's relations with others, that is, whether the person can maintain friendship, work, and social relationships. They must realize that to be successful they need to consider not only how their treatment affects the individual they are treating but, in many cases, also how the treatment affects the individual's relations with others. Furthermore, it is equally important to consider how a person's relations with others may indicate his or her need for treatment. Most instruments used to diagnose disability assess a person's social functioning, as we will see. To treat people successfully, the context in which they function must be taken into account.

The picture gets bigger. Not only are individuals treated for what ails them; groups are treated as well. Family therapy is used to heal family relations, programs are implemented to rehabilitate inner cities, and resources are allocated to help underdeveloped countries. Efforts to improve the human condition proceed on many interrelated levels. Body parts are treated when they fail to function. Individuals are treated when they can't function in society. Social groups are treated to improve their functioning so that the people in the groups might lead better lives and be less of a menace to their neighbors. The physical environment is treated to reduce pollution and to yield more raw material for people to work with.

It becomes evident that many factors must be taken into account in defining quality of life as a field of study. This requires that quality of life be viewed holistically in all of its complexity and with the recognition that a single heinous problem, such as excruciating pain, can destroy quality of life. In contrast, people who are engrossed in pursuing a single cause that is more dear to them than life may be fulfilled despite being deprived of material wealth, personal security, ample food, or housing and suffering from stress and pain.

According to the dictionary, *holism* is the view that the whole has an independent reality that cannot be understood simply through the understanding of its parts. *Holistic* is defined as being concerned with integrated systems rather than with their parts.

The parts of the prescribed holistic approach are developed as the book progresses. In Chapter 7 the parts are integrated as the holistic approach to improving the quality of life is presented. The need for a holistic approach to quality of life has been recognized. So far, nothing much has been done to prescribe one. The holistic approach developed in the book merits consideration.

THE CHALLENGE TO SCIENCE

Researchers are becoming more and more involved in studying quality of life issues. Since the middle of the twentieth century, there has been a notable increase in the amount of time, effort, and resources that have been concentrated on quality of life inquiry. The U.S. Centers for Disease Control and Prevention are conducting over a million interviews to determine the quality of life in America. Two fundamental questions are being addressed: What are the attributes or ingredients of quality of life? and, How can the quality of life be improved?

To satisfy the demand for research on quality of life, more and more instruments are being developed to assess quality of life, many of which are presented in Chapter 2. The instruments are being used to diagnose deficiencies in quality of life, to test the effectiveness of treatments designed to improve quality of life, and to monitor progress toward improvement. The instruments are used to assess the quality of life of individuals, including people suffering from disease, disability, distress, and poverty, and to assess the places people live, including communities, cities, and nations.

Despite the many instruments that have been constructed to assess quality of life, controversy abounds. Some think quality of life can be assessed only by obtaining the subjective opinion of people. Others believe quality of life can be objectively observed. Moreover, there are vast differences of opinion on the indicators and ingredients of quality of life. However, these differences need not prevent progress.

Prior research on quality of life indicators is reviewed in Chapter 2 as a basis for integrating the diverse findings into a comprehensive corpus of knowledge. This enables us to circumscribe and define quality of life as a field of study in Chapter 3, giving us a holistic view of the field. As a result the various professionals working to improve the quality of life can gain a clearer understanding of their work in context. They will be better able to relate their work to the work of others and to evaluate their contributions and the contributions of others to the overall quality of people's lives.

Yet another challenge must be met. Applications of traditional scientific method that have increased our understanding of physical matter and lower forms of life are not nearly as productive in increasing our understanding of humans and improving the quality of human life. If we are to accelerate progress in improving the quality of life, the traditional scientific method needs to be adapted for that purpose.

The teaching of the traditional scientific method has not varied appreciably since its inception. It is fundamentally a hypothesis-testing method designed to increase knowledge. It is not designed specifically to achieve improvements, including improvements in quality of life. Recommendations are made in Chapters 4 and 7 for refining traditional scientific method so that it might be more useful in improving the quality of human life.

Furthermore, in scientific inquiry, there is a need to take into account and work with human factors that affect quality of life. Attempts to improve the human condition have failed in the past because potent forces of human nature that can be worked with to improve the quality of life have been ignored. In Chapters 5, 6, and 7, major psychological factors that can, and should be dealt with in scientific research are described, and means of working with them are prescribed.

It must be acknowledged that there are approaches to improving the quality of life other than science, such as theological, philosophical, and humanistic approaches, which are not dealt with here. Different methods are used in different approaches. Drifting from one approach to another can compromise the effectiveness of inquiry. Consequently, every effort will be made to see that the proposed scientific strategy remains within the bounds of scientific inquiry. Despite its limitations, science has in the past contributed to raising the standard of living, and if adapted for the purpose, science can be applied constructively to improving the quality of life. In Chapter 4 a scientific strategy is proposed that is designed to improve the quality of life.

No other work could be found that identifies quality of life as a field of study, proposes a scientific strategy explicitly designed to improve the quality of life, and recommends ways of working with potent psychological factors to enhance success.

Basic Issues and Challenges

Passion for life is instinctive and primitive. Concern about the quality of life is to a great extent learned and rational. It arises from what humans learn about life, death, incapacitation, suffering, health, and success and from the realization that they need to know more about the quality of life to make decisions. And although during the decision-making process they may be prompted by instinctive impulses, in the main, quality of life decisions are arrived at through deliberating alternatives, predicting the consequences of following the alternatives, and selecting a preferred alternative to act on. The preferred selection may be a happy one or the lesser of a number of evils. It may concern the choice of a treatment plan, the selection of a retirement residence, or a career choice. Whatever the quality of life choice may be, it is to a great extent made rationally through the deliberation of alternatives.

The need to know more about quality of life has prompted significant research. In the next chapter, quality of life research is reviewed, providing insights into the state of the science and a foundation for improving the quality of life in the future. The state of the science is quite limited. The outlook is promising.

Chapter 2

Data to Build Upon

The development of a holistic scientific strategy begins with a review of research on quality of life indicators. This review will serve as a basis for identifying essential ingredients of quality of life and defining quality of life as a field of study in the next chapter. There is a pressing need to define quality of life as a field of study so that quality of life can be studied and enhanced more expeditiously. At present, there is a great deal of ambiguity and controversy on quality of life, elements, and indicators. Rather than argue about these issues, let's see what the research tells us.

The review will be limited to research on quality of life indicators. There is considerably more research on other aspects of quality of life, for example, the effects of various diseases and treatments on quality of life. But this is beyond the scope of our concern in this chapter. Furthermore, it is necessary to define and assess quality of life more accurately before we can effectively evaluate the impact of treatments on quality of life. The survey reveals mainly definitions of quality of life indicators, methods of observing them, and actual observations of the indicators in a wide variety of settings.

Perhaps the clearest way to present indicators is by using tables displaying and explicating them as well as writing about them. This provides the opportunity to see in a table the range of indicators used for a particular purpose and to compare tables to see the differences in the indicators used for different purposes. References are given so that more detail can be obtained.

Two approaches have been taken to identify quality of life indicators. Societal and personal indicators of quality of life have, for the most part, been distinguished and studied separately. We will consider societal indicators and then personal indicators. Since the essential ingredients of quality of life have yet to be identified to any great extent, the following presentation will consist mainly of quality of life indicators and instruments rather than confirmed conclusions.

SOCIETAL INDICATORS OF QUALITY OF LIFE

Societies, that is, groups of people who share common languages, rituals, and beliefs, have been studied to determine the quality of life in particular locations: communities, cities, or nations. Societies are of concern because, as we know, where people happen to live affects the quality of their lives. The quality of life in societies is often studied to determine where to allocate resources. One implication is that if resources are allocated to a needy society, the lives of the people in it can be improved. It's also considered to be a way to placate frustrated, belligerent societies who are threats to others. The extent to which these goals are achieved has still not been resolved, but social scientists are intently studying the problems. A second reason for studying the quality of life in societies is to determine the best places to live. For example, surveys are conducted to identify the best places to retire or the best places to locate an industry. Societal studies report normative data indicating the general living conditions in a location—for example, average life expectancy, unemployment rate, and crime rate.

The data collected in studies of the quality of life in societies vary considerably, as we will see. Still, there has been some agreement that data need to be collected in six broad categories (Smith 1973, 70). And data are collected in these categories in a great many studies. The following example familiarizes us with basic attributes of quality of life in societies. Table 2.1 shows the six general categories and the data that were collected in each category to study the quality of life in American cities (Schneider 1976).

An interesting recent study of societal indicators has been conducted at Fordham University's Institute for Innovation in Social Policy (Miringoff 1995). Conductors of the study use an index of social health to monitor the social well-being of the United States that is somewhat different, and they make significant comparisons between trends in their index and other trends. The index combines in one measure the following sixteen social problems:

Children: Infant mortality
 Child abuse
 Children in poverty
Youth: Teen suicide
 Drug abuse
 High school dropouts
Adults: Unemployment
 Average weekly earnings
 Health insurance coverage

Table 2.1
Indicators of the Quality of Life in American Cities

I. Income, wealth, and employment

    a. Percent of labor force unemployed

    b. Percent of households with income less than $3,000

    c. Per capita income

II. Environment

    a. Percent substandard dwellings

    b. Air quality (average yearly concentration of three air pollution components)

    c. Cost of transportation for a family of four

III. Health

    a. Infant (under one year) deaths per 1,000 live births

    b. Reported suicide rates per 100,000

IV. Education

    a. Median school years completed by adult population

V. Participation and alienation

    a. Percent of voting-age population that voted in presidential election

    b. Per capita contribution to United Fund Appeal

VI. Social disorganization

    a. Reported robberies per 100,000

    b. Reported narcotics addiction rate

Source: Schneider 1976, 301.

Figure 2.1
Index of Social Health and Gross Domestic Product, 1970–1993

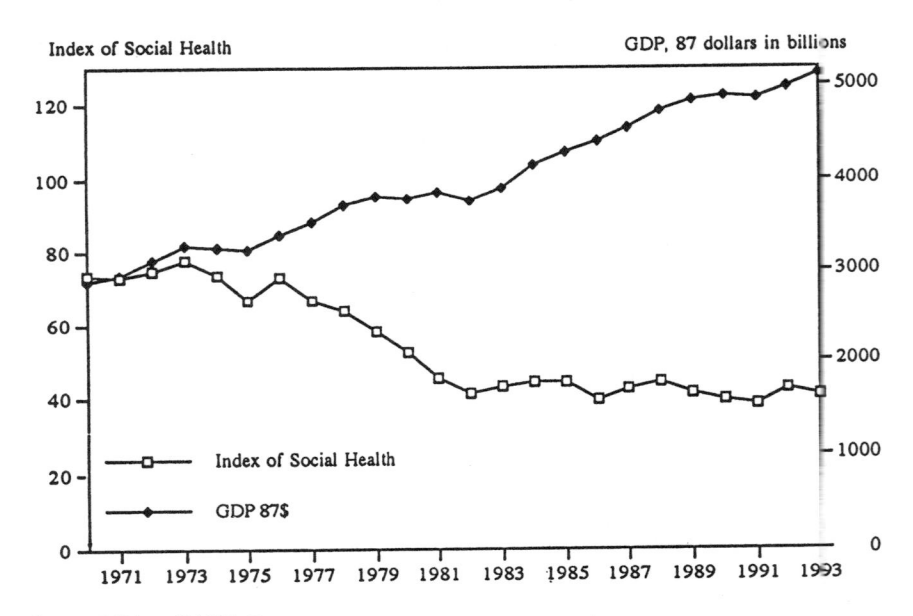

[Line graph showing Index of Social Health declining from about 75 in 1970 to about 45 in 1993, while GDP (87$ in billions) rises steadily from about 3000 to over 5000 over the same period.]

Source: Miringoff 1995, 7.

Aging: Poverty among those over 65
              Out-of-pocket health costs for those over 65
All ages: Homicides
              Alcohol-related traffic fatalities
              Food stamp coverage
              Access to affordable housing
              Gap between rich and poor

Figure 2.1 shows trends in the Index of Social Health and compares them to trends in the gross domestic product (GDP) of the United States, which is commonly used to indicate economic status and growth.

The conductors of the study conclude that

the overall trends reflected by the Index of Social Health are cause for concern. The Index has dropped steadily since 1970, and did so again in 1993, declining to the fourth worst point in the twenty-four year period. That our social health has fallen so far below what we achieved in the early and middle seventies, and that it has remained at so low a level for so long, has serious implications for American society. The deep and persistent decline of conditions affecting the health, income, and physical security of Americans across the age spectrum requires particular attention.

Table 2.2
Money **Magazine's Top Ten Metropolitan Areas to Live in, in the United States**

		Health	Crime	Economy	Housing	Education	Transit	Weather	Leisure	Arts
1.	Raleigh/Durham/Chapel Hill, N.C.	88	20	93	88	95	41	38	4	28
2.	Rochester, Minn.	96	63	71	43	97	81	14	25	26
3.	Provo/Orem, Utah	59	58	79	61	41	58	29	37	22
4.	Salt Lake City/Ogden	72	27	81	75	51	40	26	35	22
5.	San Jose	82	35	37	25	40	27	83	93	91
6.	Stamford/Norwalk, Conn.	86	53	74	40	14	18	28	88	100
7.	Gainesville, Fla.	45	4	92	49	45	45	79	5	22
8.	Seattle	79	20	67	28	62	28	47	94	52
9.	Sioux Falls, S.D.	71	54	96	43	6	75	9	2	14
10.	Albuquerque	61	13	63	94	36	42	43	23	17

Note: Metropolitan areas are listed in rank order from 1 (best) to 10. Higher scores under particular factors indicate better living conditions.
Source: "Best Places to Live in America," *Money*, September 1994, p. 132.

Equally significant is the change in the relationship between overall economic growth, as measured by the Gross Domestic Product, and social health. Until the mid-1970s, growth and social health demonstrated similar patterns. After that time there was a striking change; social health has declined and GDP has continued to rise. This divergence carries important implications for both public policy and helps to clarify the changes taking place in American society. (1995, 10)

Efforts aimed solely at stimulating overall economic growth may not have the desired effect on many of the indicators of social health, as once was the case.

Less academic studies are conducted by popular magazines to appeal to their readers. *U.S. News and World Report* conducts periodic surveys that distinguish among the living conditions in metropolitan areas. (A report appears in its April 11, 1994, issues [see "Home Guide," pp. 57–83].) And *Fortune* magazine conducts surveys to determine the best metropolitan areas in which to do business.

Table 2.2 shows the nine indicators used by *Money* magazine in its September 1994 issue, "Best Places Live in America," to rank 300 metropolitan areas in the United States (p. 132). The rank order of the top ten metropolitan areas is also shown. A survey to determine the interests of its readers is conducted periodically as a basis for selecting the indicators that will be used, and the indicators vary from time to time. The rankings are not established for the purpose of making a particular decision, such as deciding where to locate an industry. Notice, too, that *Money* magazine's survey includes indicators not included in Table 2.1 or in the Index of Social Health, specifically arts, weather, and leisure.

The Raleigh-Durham–Chapel Hill, North Carolina, area ranked as the best metropolitan area to live in, in the United States. It ranked especially high in the economic, housing, education, and health categories. The region has an unemployment rate of 3 percent, three large universities, and large medical centers. The bottom ten areas were Yuba City, California; Springfield, Illinois; Toledo, Ohio; Peoria, Illinois; Detroit, Michigan; Glens Falls, New York; Flint, Michigan; Saginaw–Bay City–Midland, Michigan; Rockford, Illinois; and Jackson, Michigan. Jackson, a city of 150,000, ranked at the bottom partly because it has the state's largest prison. And crime in the prison is recorded as crime in Jackson.

In 1984 the United States Centers for Disease Control and Prevention (CDC) embarked on an extensive telephone survey (called the Behavioral Risk Factor Surveillance System) to determine the health status of noninstitutionalized people in the United States ages eighteen and older. The new mission of the CDC is to promote health and quality of life by preventing and controlling disease, injury, and disability. Improving quality of life has become a primary emphasis. A main objective is to increase the span of healthy life for all persons in the United States. The CDC is also interested in identifying groups of people with relatively poor health and quality of life so that it might intervene to improve their status. Table 2.3 contrasts how people perceive their health in 49 of the 50 states.

The results were obtained in a 1993 survey from 102,263 respondents in answer to the question: Would you say that your health is excellent, very good, good, fair, or poor? (Centers for Disease Control and Prevention [CDC] 1995. The entire questionnaire is presented in Table 2.11, and additional discussions of their surveys are presented when personal quality of life indicators are considered.

Somewhat different societal quality of life indicators were used to study the quality of life in nations (Slottje et al. 1991). These indicators show how the United States ranks among 126 nations on each of the indicators (Table 2.4). Despite the leadership role of the United States among nations, it does not rank consistently high on important quality of life attributes.

Table 2.4 provides an example of a study of the relative strengths and weaknesses of the quality of life in one society, the United States. Table 2.5 provides a comparative example of the quality of life in different societies. The rankings of 40 nations on the same societal quality of life indicators combined are shown.

The inclusion of countries like Switzerland, the United Kingdom, New Zealand, Sweden, Canada, the United States, and Japan in the top 20 is not much of a surprise. The presence of countries like New Guinea, Jamaica, Barbados, and Gambia are somewhat more surprising. Had the rankings relied solely on one dimension, like real gross domestic product (RGDP), for example, none of these countries would be in the top 30. The other criteria give a different di-

Table 2.3
Where They Feel Best

STATE		STATE	
Alaska	91.6	Pennsylvania	87.2
New Hampshire	90.8	Delaware	87.2
New Jersey	90.8	Nevada	87.0
Iowa	90.7	California	87.0
Washington	90.5	Michigan	87.0
Connecticut	90.0	Montana	86.5
Dist. of Columbia	89.9	North Dakota	86.4
Minnesota	89.6	Ohio	86.3
Massachusetts	89.6	Georgia	86.3
Vermont	89.5	Hawaii	86.2
South Dakota	89.4	Texas	86.0
Maryland	89.2	Indiana	85.3
Wisconsin	89.0	Alabama	85.3
Virginia	89.0	Rhode Island	85.2
Colorado	88.9	Florida	84.6
Utah	88.1	Louisiana	83.9
Idaho	88.1	Missouri	83.8
Nebraska	88.0	Oklahoma	82.9
Kansas	88.0	North Carolina	82.8
Oregon	88.0	South Carolina	82.7
New York	87.6	Tennessee	81.5
Arizona	87.5	Arkansas	80.3
New Mexico	87.4	Kentucky	79.9
Maine	87.4	Mississippi	78.4
Illinois	87.4	West Virginia	76.6

Note: States are rank ordered in accordance with the derived scores shown.
Source: Centers for Disease Control and Prevention (CDC) 1995.

Table 2.4
U.S. Ranking among 126 Nations on 20 Quality of Life Indicators

U.S. Ranking	Indicators
14th	Political rights (e.g., legitimate election of government officials, campaigning opportunities, minority self-determination)
10th	Civil liberty (e.g., freedom of press, speech, assembly, religion)
2d	Average household size
108th	Soldiers per 1,000 civilians
3d	Energy consumption per capita
37th	Percent of women in labor force
30th	Percent of children in labor force
27th	Length of roads per square kilometer of territory
1st	Telephones per capita
19th	Male life expectancy
9th	Female life expectancy
17th	Infant mortality rate per 1,000 births
32d	Population per hospital bed
20th	Population per physician
4th	Daily calorie consumption
13th	Male literacy rate
11th	Female literacy rate
1st	Radio receivers per 1,000 people
2d	Number of daily newspapers
1st	Real gross domestic product (adjusted gross national product)

Note: Higher ranking indicates higher quality of life.
Source: Slottje et al. 1991.

Table 2.5
Quality of Life Rankings of Nations on the 20 Quality of Life Indicators (in Table 2.4) Combined

1st	Switzerland	21st	Denmark
2nd	United Kingdom	22nd	Botswana
3rd	New Zealand	23rd	Hong Kong
4th	Jamaica	24th	Senegal
5th	New Guinea	25th	Honduras
6th	Canada	26th	Uruguay
7th	Austria	27th	Netherlands
8th	Luxembourg	28th	Finland
9th	Australia	29th	Norway
10th	Sweden	30th	Dominica
11th	Mauritius	31st	Bolivia
12th	Barbados	32nd	Italy
13th	United States	33rd	Fiji
14th	Iceland	34th	Kenya
15th	Japan	35th	Trinidad-Tobago
16th	Gambia	36th	Belgium
17th	Costa Rica	37th	Spain
18th	Portugal	38th	Uganda
19th	Ireland	39th	Argentina
20th	Ghana	40th	Colombia

Source: Slottje et al. 1991.

mension to the rankings. By incorporating liberty indicators into the analysis, countries known for political repression finish at the bottom of the rankings even though some of them have relatively higher RGDP and physical quality of life attributes like life expectancy. It may also be surprising to find that the United States ranks 13th on the combination of quality of life attributes.

In the following quality of life analysis, the United States ranks fourth among members of the Organization for Economic Cooperation and Development (OECD) nations when a different combination of quality of life indicators is used (Scheer 1980) (see Table 2.6). The difference in rankings in Table 2.6 and Tables 2.4 and Table 2.5 is due in large measure to the differences in the quality of life indicators used. In general, the quality of life indicators in Tables 2.4 and 2.5 are more comprehensive, placing greater emphasis on political factors such as political rights, civil liberty, soldiers per 1,000 civilians. As Table 2.4 shows, the United States does not rank as high on these factors as on other factors.

The United States is a pluralist melting pot. Clashes among social factions and haves and have-nots are commonplace and seem to generate continuing upheaval, legal battles, and violence. The civil liberties guaranteed in the Constitution are far from being realized in daily life. Moreover, the United States has developed the most technically advanced war machines to protect themselves from foreign enemies and to lend muscle to United Nations' edicts when the United States agrees with them—while internally crime runs rampant. In addition, although the United States has developed more advanced medical procedures, available medical treatment is not delivered to as large a percentage of the total population as in other countries. As a result, the United States does not rank at the top on indicators such as infant mortality or hospital beds and physicians per population. It is no wonder that crime, warfare, and health care hang heavy on the minds of the public.

It appears that the United States obtains its highest ranking on criteria relating to riches. In December 1994 the World Bank reported that the United States is the seventh richest nation, as measured by the economic output per person. The top ten rankings were as follows.

1. Switzerland $36,410
2. Luxembourg   35,850
3. Japan   31,450
4. Denmark   26,510
5. Norway   26,340
6. Sweden   24,830
7. United States 24,750
8. Iceland   23,620

Table 2.6
Quality of Life Comparison of OECD Nations

Rank	Nation	Quality of Life Indicators
1	Sweden	1. Per capita GNP increase in constant prices
2	Switzerland	2. Life expectancy of a newborn boy
3	Norway	3. Life expectancy of a newborn girl
4	United States	4. Life expectancy of a 40-year-old man
5	Canada	5. Life expectancy of a 40-year-old woman
6	Denmark	6. Life expectancy of a 60-year-old man
7	Netherlands	7. Life expectancy of a 60-year-old woman
8	Japan	8. Excess consumption of calories
9	France	9. Physicians per population
10	United Kingdom	10. Child mortality (2nd to 12th month) per live births
11	Belgium	11. Fatal work accidents (average for sectors included)
12	Federal Germany	a in mining
13	Austria	b in manufacturing
14	Finland	c in railways
15	Italy	d in construction

12. Homicides per population
13. Fatal traffic accidents per population
14. Expenditures for food and tobacco as percentage of total spending
15. Discretionary spending as percent of total (included here are total consumer expenditures after rent, heat, light, food, clothing and shoes have been deducted)
16. Private cars per population
17. Average number of persons per room
18. a Percentage of dwelling units with running water
    b Percentage of dwelling units with bathroom
19. Telephones per population
20. TV sets per population
21. Primary schoolchildren per teacher
22. Women as percentage of university-level students
23. Change in the ratio of economically active persons to population of working age
24. Unemployment as percentage of labor force
25. Average work week of full-time workers in manufacturing in hours

Source: Scheer 1980.

Table 2.7
Indicators of Quality of Life: United States versus Western Europe

	United States	Western Europe
Fresh fruit as a percent of all fruit consumed	62%	87.2%
Fresh, frozen, or smoked meats as a percent of all meat consumed	66%	90.5%
Average minutes per day devoted to walking, hiking, playing outdoors, and engaging in active sports	7.9 min.	28.5 min.
Percentages of adult population taking annual vacations of six or more days	27.7%	44%
Average time spent alone while awake per day	6.6 hrs.	4.1 hrs.
Average minutes per day devoted to gardening and pets	3.3 min.	16.8 min.
Expenditures on flowers as a percent of national income	0.20%	0.39%

Source: Scitovsky 1976.

| 9. Germany | 23,560 |
| 10. Kuwait | 23,350 |

The United States ranks second after adjusting for differing price levels among nations to reflect buying power. Luxembourg had the highest real output per person at $29,510, followed by the United States at $24,750. Costlier goods and services reduce consumers' buying power, lowering the real per capita income in wealthier nations.

Tibor Scitovsky (1976) in his book *The Joyless Economy* presents the comparisons in Table 2.7 to show that the quality of life in the United States is not as high as in Western Europe. The quality of life indicators Scitovsky uses appear to be limited to his ideas of healthful practices.

PERSONAL QUALITY OF LIFE INDICATORS

Personal quality of life indicators have been used to determine how to allocate resources to improve health care, to assess the effects of treatments, and to help people fulfill their aspirations, among other things.

When the medical profession moved beyond merely treating diseases to directing the rehabilitation of patients' functional disabilities as well, they became interested in finding methods of assessing functional ability. Doctors were traditionally trained to observe patient progress primarily by noting the reduction of disease symptoms. In many cases, it is not difficult to tell when doctors are successful in treating diseases and ailments of body organs. For instance, it is not difficult to determine when a patient's temperature returns to normal, a rash disappears, an X-ray shows that a bone has mended, the swelling of a sprain subsides, or allergic symptoms disappear.

It is much more difficult for doctors to determine patients' functional abilities and distress. They were, and still are, in need of instruments that will enable them to determine how well people are able to perform vital functions and their attitudes toward life. Several instruments have been constructed. Some are observation instruments, while others are self-report instruments. The instruments are used more for clinical purposes such as diagnosing personal difficulties, assessing progress toward recovery, and allocating resources. They are used less often for basic research, which partially explains why we are not able to generalize more about personal indicators of quality of life.

Observation Instruments

Observation instruments are used by qualified observers such as clinicians and interviewers to assess personal characteristics of people. When the use of an observation instrument requires specialized knowledge or skills, training is required to ensure the accuracy of observations.

The Health Status Index, constructed by Milton Chen and others, is an observation instrument used by qualified observers to distinguish functional levels of people (Brock 1993) (see Table 2.8). As one can see, functional disability is not nearly as easy to observe as many disease symptoms such as body temperature.

The expanded focus in the medical profession on the rehabilitation of functional disabilities has impacted health care movements throughout the world. The allocation of health care resources is no longer limited to the treatment of disease and organ ailments. The trend is to allocate resources to observe the effects of treatments on restoring function so that money can be provided for the treatments that are the most successful.

At first, life expectancy was considered for use as an index of treatment effectiveness. But the idea was abandoned because it is difficult, if not impossible, to determine the extent to which a treatment extends life. In addition, there

Table 2.8
Scales and Definitions for the Classification of Function Levels

Scale	Step	Definition
		Mobility scale
5	Travelled freely	Used public transportation or drove alone. For below 6 age group, travelled as usual for age.
4	Travelled with difficulty	(a) Went outside alone, but had trouble getting around community freely, or (b) required assistance to use public transportation or automobile.
3	In house	(a) All day, because of illness or condition, or (b) needed human assistance to go outside.
2	In hospital	Not only general hospital, but also nursing home, extended care facility, sanatorium, or similar institution.
1	In special unit	For some part of the day in a restricted area of the hospital such as intensive care, operating room, recovery room, isolation ward, or similar unit.
0	Death	
		Physical activity scale
4	Walked freely	With no limitations of any kind.
3	Walked with limitations	(a) With cane, crutches, or mechanical aid, or (b) limited in lifting, stooping, or using stairs or inclines, or (c) limited in speed or distance by general physical condition.
2	Moved independently in wheelchair	Propelled self alone in wheelchair.
1	In bed or chair	For most or all of the day.
0	Death	
		Social activity scale
5	Performed major and other activities	*Major* means specifically: play for below 6, school for 6–17, and work or maintain household for adults. *Other* means all activities not classified as major, such as athletics, clubs, shopping, church, hobbies, civic projects, or games as appropriate for age.
4	Performed major but limited in other activities	Played, went to school, worked, or kept house but limited in other activities as defined above.
3	Performed major activity with limitation	Limited in the amount or kind of major activity performed, for instance, needed special rest periods, special school, or special working aids.
2	Did not perform major activity but performed self-care activities	Did not play, go to school, work or keep house, but dressed, bathed, and fed self.
1	Required assistance with self-care activities	Required human help with one or more of the following — dressing, bathing, or eating — and did not perform major or other activities. For below 6 age group, means assistance not usually required for age.
0	Death	

Source: Brock 1993.

Table 2.9
Matrix of Health Statistics

	Distress (physical and mental)			
Disability	No distress	Mild	Moderate	Severe
1				
2				
3				
4				
5				
6				
7				
8				

1 = No disability.
2 = Slight social disability.
3 = Severe social disability and/or slight impairment of performance at work. Able to do all housework except very heavy tasks.
4 = Choice of work or performance at work very severely limited. Housewives and old people able to do light housework only but able to go out shopping.
5 = Unable to undertake any paid employment. Unable to continue any education. Old people confined to home except for escorted outings and short walks and unable to do shopping.
6 = Confined to chair or to wheelchair or able to move around in the home only with support from an assistant.
7 = Confined to bed.
8 = Unconscious.
Source: Rosser and Kind 1978, 349.

was an interest in improving functional ability while people are alive, as well as extending life span. This gave rise to the development of a quality of life instrument that proposes to take functional ability and life expectancy into account, the Quality Adjusted Life Years, or QUALYs, Index. The observation instrument shown in Table 2.9 was conceived to provide data needed to calculate QUALYs (Rosser and Kind 1978).

As shown, there are two dimensions to the matrix: (1) disability, which, like the instrument in Table 2.8, assesses functional ability, and (2) distress, which attempts to assess the pain and suffering of the patient. Distress is assessed presumably to provide a more complete picture of quality of life. Observers score patients on both dimensions to determine the state of their illnesses. The results are used in a formula to compute QUALYs. The QUALYs are then used to assess treatment effectiveness (Kind, Claire, and Godfrey 1990). The more a treatment increases

Quality Adjusted Life Years, the more effective the treatment is considered to be—and the more likely it is that the treatment will receive support.

QUALYs assess the benefits of health care treatments, enabling economists to do cost/benefit analyses for agencies that fund the treatments. The use of QUALYs was developed at the University of York in England to assist in solving the resource shortage problems often attendant to socialized medicine. However, there is a need to assess treatment effectiveness anywhere when new treatments are being introduced that propose to remedy previously uncontrolled dire illnesses. Furthermore, there will never be unlimited government funds anywhere to pay for all the treatments people might want or need to try. And even when cost is not a consideration, treatment effects need to be assessed in order to approve treatments for public use (see Gelber, Gelman, and Goldhirsch 1989; Patrick and Erickson 1993.)

The medical profession and the agencies that authorize and support medical treatments took a giant step forward when they expanded their focus and responsibility from curing disease to rehabilitating functional disabilities as well. However, they have not gone far enough. They need to further extend their focus to include the promotion of superior functional abilities. The indicators on the instruments discussed range only from extreme disability, for instance, death or unconsciousness, to no disability. There are no indicators on the instruments of superior functional ability. Yet superior functional ability is an important sign of superior health. And attainment of superior health should be a health care goal. We need to do more than reduce disability; we should strive to extend people's capabilities. A complete functional ability scale would range from extreme disability to no disability to superior functional ability:

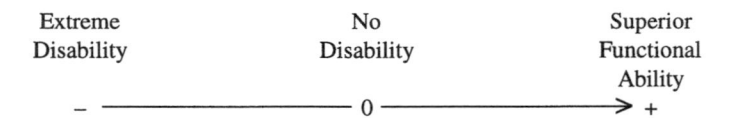

Extreme Disability	No Disability	Superior Functional Ability
— ——————————	0 ——————————	→ +

Moreover, the meaning of functional ability needs to be reconsidered in order to refine scales that propose to assess functional ability. Since people don t live in isolation, functional ability, in essence, must have to do with people's ability to interact with their environments, including other people. A scale constructed to assess functional ability should range from inability to interact with the environment to superior ability to interact with the environment. Just assessing a person's independence or ability to perform physical feats, such as walking or weight lifting, is not sufficient.

A "distress" scale also presents problems as an index of health on an observation instrument. It is much less difficult for observers to assess functional ability than to assess distress. Functional ability can be determined by observing how people interact with their environments, for example, their ability to cook, shop, and feed and dress themselves. It is much more difficult to determine people's distress by observing them. Does the observer note facial expressions or complaints to derive a distress rating? Or does the observer ask people to rate their own distress on a scale, in which case the rating is a self-report rather than the rating of an observer.

Self-Report Instruments

Quality of life data are obtained from self-report instruments as well as from observation instruments. Not only are observers able to assess the quality of people's lives; people can reveal information on the quality of their lives in interviews, on questionnaires, and on testing instruments.

Angus Campbell and his associates at the Institute for Social Research, University of Michigan, were among the first to recognize the importance of personal responses to quality of life issues. He pointed out the limitations of economic and social indicators that are usually derived from government statistics. In their seminal work, he and his colleagues demonstrated that personal indicators of quality of life differ substantially from economic and social indicators (see Campbell 1976; Campbell, Converse, and Rodgers 1976). Schneider's (1976) findings were similar. However, they were less successful in deriving personal indicators that account for the quality of people's lives. In the Institute for Social Research study, 2,164 people from 48 states were interviewed for an average of one hour and fifteen minutes following a structured questionnaire that probed for satisfaction with life, the affective quality of life, and perceived stress. The structured questionnaire (about 40 pages) is too extensive to show here (but see Campbell, Converse, and Rodgers 1976). However, much of the same terrain is covered in some of the following tables.

Prior discussion of instruments used to assess states of health describes instruments used by observers such as doctors and nurses to assess disability and distress in individuals. There are self-report instruments that assess states of health as well, based on subject responses. The Nottingham Health Profile developed at Nottingham University, England, is a self-report questionnaire in two sections. Section 1 probes for information on physical mobility, pain, energy, sleep, emotional reactions, and social isolation. Section 2 asks individuals whether their present state of health is causing problems with various parts of their lives such as their social lives, work, home lives, or sex lives (Nordenfeldt 1993, 115). Another self-report instrument used to assess states of health is the Sickness Impact Profile (SIP) (Brock 1993) (see Table 2.10). The items are slightly different than the other functional ability instruments.

To implement its quality of life surveillance program, the Centers for Disease Control and Prevention developed Behavioral Risk Factor Surveillance System (BRFSS) questionnaires. The responses to the questionnaires are to "provide a basis for projecting the demands for health services, developing targeted intervention programs, allocating resources, and evaluating intervention effects" (Hennessey et al. 1994). As indicated, the responses are obtained by phoning U.S. noninstitutionalized Americans ages eighteen and older. The first, the Core

Table 2.10
Sickness Impact Profile Categories and Selected Items

Dimension	Category	Items describing behavior related to:	Selected items
Independent categories	SR	Sleep and rest	I sit during much of the day I sleep or nap during the day
	E	Eating	I am eating no food at all; nutrition is taken through tubes or intravenous fluids I am eating special or different food
	W	Work	I am not working at all I often act irritable toward my work associates
	HM	Home management	I am not doing any of the maintenance or repair work around the house that I usually do I am not doing heavy work around the house
	RP	Recreation and pastimes	I am going out for entertainment less I am not doing any of my usual physical recreation or activities
I. Physical	A	Ambulation	I walk shorter distances or stop to rest often I do not walk at all

	M	Mobility	I stay within one room
			I stay away from home only for brief periods of time
	BCM	Body care and movement	I do not bathe myself at all but am bathed by someone else
			I am very clumsy in my body movements
II. Psychosocial	SI	Social interaction	I am doing fewer social activities with groups of people
			I isolate myself as much as I can from the rest of the family
	AB	Alertness behavior	I have difficulty reasoning and solving problems, e.g., making plans, making decisions, learning new things
			I sometimes behave as if I were confused or disoriented in place or time, e.g., where I am, who is around, directions, what day it is
	EB	Emotional behavior	I laugh or cry suddenly
			I act irritable and impatient with myself, e.g., talk badly about myself, swear at myself, blame myself for things that happen
	C	Communication	I am having trouble writing or typing
			I do not speak clearly when I am under stress

Source: Brock 1993, 120–121.

Table 2.11
Quality of Life Questions Now on the Core BRFSS Questionnaire

The interview will only take a short time, and all the information obtained in this study will be confidential.

Section 1: Health Status

1. Would you say that in general your health is: (33)

   Please Read
   a. Excellent   1
   b. Very good   2
   c. Good   3
   d. Fair   4
              or
   e. Poor   5

   Do not read these responses
   Don't know/Not sure   7
   Refused   9

2. Now thinking about your physical health, which includes physical illness and injury, for how many days during the past 30 days was your physical health not good? (34-35)
    a. Number of days ___
    b. None   8 8
       Don't know/Not sure   7 7
       Refused   9 9

3. Now thinking about your mental health, which includes stress, depression, and problems with emotions, for how many days during the past 30 days was your mental health not good? (36-37)
    a. Number of days ___
    b. None **If Q. 2 also "None," skip next question**   8 8
       Don't know/Not sure   7 7
       Refused   9 9

4. During the past 30 days, for about how many days did poor physical or mental health keep you from doing your usual activities, such as self-care, work, or recreation? (38-39)
    a. Number of days ___
    b. None   8 8
       Don't know/Not sure   7 7
       Refused   9 9

Source: CDC 1994.

questionnaire, assesses health status (see Table 2.11). The second, the Optional Module, and the third, the Unmet Needs Module, have been developed more recently. To date, more than 240,000 people have been asked these questions, primarily the Core questions, and well over 1 million interviews are projected by the year 2000.

Responses from 44,978 persons reveal that in 21 states 15% percent of respondents reported "fair" or "poor" health; 32%, recent physical health limitations; 31%, recent mental health limitations; and 19%, recent activity limitations. Good health days (GHDs) were also calculated by subtracting the sum of "not good" physical health days and "not good" mental health days from 30 days (with the restriction that the number of GHDs cannot be less than zero).

Of the characteristics studied, the mean number of GHD during the 30 days preceding the survey was highest for persons with annual household incomes of more than $50,000 (26.4 days), college graduates (26.2), and Asians/Pacific Islanders (26.2). The mean number of GHD was lowest for persons who were aged 75 years (23.0), who smoked 20 or more cigarettes per day (22.9), who were told by a health professional more than once they have high blood pressure (22.1), who were unemployed (22.0), who were separated from their spouses (22.0), who had less than a high school education (21.9), who had annual household incomes of less than $10,000 (21.1), who were told by a physician they have diabetes (19.9), and who were unable to work (10.7).

Mean numbers of GHDs varied substantially when respondents were grouped by annual household income, education, age group, and sex. The mean number of GHDs was lowest (17.5 days) for men aged 35–49 years who had annual household incomes of less than $10,000 and a high school education or less (n = 167). Each of the five groups with the lowest mean number of GHDs (less than 20 days) comprised persons aged 35–64 years who had an annual household income of less than $10,000 (combined n = 362 men, 1140 women). The mean number of GHDs was highest (27.9 days) for men aged 50–64 years who had annual household incomes of more than $50,000 and at least some college education (n = 646). Each of the five groups with the highest mean number of GHDs (27 or more days) comprised men aged 35 years who had annual household incomes of more than $50,000 (combined n = 2842). (CDC 1994, 376)

In a survey of 102,263 respondents using the Core instrument and survey technique, "86.6% of respondents reported good to excellent self-rated health" (CDC 1995, 196).

The second questionnaire consists of ten functional status questions that attempt to obtain information on factors that can limit functioning (Table 2.12).

The unmet needs questionnaire was developed to determine the extent to which people are obtaining the health care they may need. Unmet health needs were considered to be a significant determinant of quality of life. Because the questionnaire is lengthy, an abridged version is presented in Table 2.13, showing only the main questions.

Little has been reported in the way of findings as yet from the administration of this questionnaire. However, some findings were reported on unmet health needs from the Older Americans Resources and Services (OARS) Project data conducted in Cleveland (Pfeiffer 1975; Fillenbaum 1988).

The most significant unmet need presented by both elders and service providers was the need for financial assistance for prescription drugs. Inability to afford adequate eye and dental care, as well as aids such as eyeglasses, hearing aids, and dentures, was also cited as an important factor. Transportation for medical appointments was another key area. (Diwan 1994)

A conceptual framework for analyzing health care needs of community-dwelling elderly has been proposed. In it, unmet needs are classified as (1) basic maintenance (food), (2) supportive (home care/meals), (3) rehabilitative (physical therapy, mental health services), (4) treatment (medical), (5) promotive (physical activity), and (6) preventive (risk factor reduction). Factors contributing to unmet needs are also described (Diwan and Moriarty 1995).

Since over 1 million telephone interviews are projected between now and the year 2000, we can expect more reports on responses to questions in the Optional Module and Unmet Needs Module from now on.

The National Center for Health Statistics, a division of the CDC, is using morbidity and mortality data to derive a measure of years of healthy life (YHL) for the purpose of tracking progress toward increasing the healthy life span of Americans (Erickson, Wilson, and Shannon 1995). First, health-related quality of life (HQL) was operationally defined using two types of information, namely, perceived health, obtained from responses to question 1 of the Core questions used in the Nation Health Interview Survey (NHIS), and role or activity limitations based on responses to other questions on the NHIS. Role limitation captures people's limitations in social roles usually associated with their age group. Each person is classified in one of six categories (Table 2.14).

The operational definition of HQL was based on a matrix of five categories of perceived health and six categories of role limitations. This results in 30 possible health states, as shown in Table 2.15, (Erickson, Wilson, and Shannon 1995). As shown in Table 2.15,

values were assigned to each of the 30 cells in the matrix. . . . Values range from 1.00 for persons who have no role limitations and are in excellent health to 0.10 for persons who are limited in Activities of Daily Living (ADL) and are in poor health. According to these values, if a person lives 1 year in excellent health and has no limitation of activity, then he or she has 1 full year of health life. Other health states result in less than a full year of health life.

"It is possible to estimate an indiviual's health-related quality of life by using the health states defined by activity limitation and perceived health along with the values for these weights."

After health-related quality of life data are derived as described, they are combined with mortality data to derive years of healthy life. In addition to being used to track progress toward increasing healthy life span, estimates of remaining years of healthy life can be compared to estimates of remaining life years

Table 2.12
Quality of Life Optional BRFSS Module

These next questions are about limitations you may have in your daily life.

1. Are you LIMITED in any way in any activities because of any impairment or health problem?
   a. Yes 1
   b. No **Go to Q. 6** 2
      Don't know/Not sure **Go to Q. 6** 7
      Refused **Go to Q. 6** 9

2. What is the MAJOR impairment or health problem that limits your activities?
   Do Not Read. Code Only One Category.
   a. Arthritis/rheumatism 01
   b. Back or neck problem 02
   c. Fractures, bone/joint injury 03
   d. Walking problem 04
   e. Lung/breathing problem 05
   f. Hearing problem 06
   g. Eye/vision problem 07
   h. Heart problem 08
   i. Stroke problem 09
   j. Hypertension/high blood pressure 10
   k. Diabetes 11
   l. Cancer 12
   m. Depression/anxiety/emotional problem 13
   n. Other impairment/problem 14
      Don't know/Not sure 77
      Refused 99

6. During the past 30 days, for about how many days did PAIN make it hard for you to do your usual activities, such as self-care, work, or recreation?
   a. Number of days __
   b. None 88
      Don't know/Not sure 77
      Refused 99

7. During the past 30 days, for about how many days have you felt SAD, BLUE, or DEPRESSED?
   a. Number of days __
   b. None 88
      Don't know/Not sure 77
      Refused 99

8. During the past 30 days, for about how many days have you felt WORRIED, TENSE, or ANXIOUS?
   a. Number of days __
   b. None 88
      Don't know/Not sure 77
      Refused 99

9. During the past 30 days, for about how many days have you felt you did NOT get ENOUGH REST or SLEEP?
   a. Number of days __
   b. None 88

3. For HOW LONG have your activities been limited because of your major impairment or health problem?
   Do Not Read. Code using respondent's unit of time.
   a. Days   1 _ _
   b. Weeks   2 _ _
   c. Months   3 _ _
   d. Years   4 _ _
       Don't know/Not sure   7 7 7
       Refused   9 9 9

4. Because of any impairment or health problem, do you need the help of other persons with your PERSONAL CARE needs, such as eating, bathing, dressing, or getting around the house?
   a. Yes   1
   b. No   2
       Don't know/Not sure   7
       Refused   9

5. Because of any impairment or health problem, do you need the help of other persons in handling your ROUTINE needs, such as everyday household chores, doing necessary business, shopping, or getting around for other purposes?
   a. Yes   1
   b. No   2
       Don't know/Not sure   7
       Refused   9

       Don't know/Not sure   7 7
       Refused   9 9

10. During the past 30 days, for about how many days have you felt VERY HEALTHY AND FULL OF ENERGY?
    a. Number of days   _ _
    b. None   8 8
        Don't know/Not sure   7 7
        Refused   9 9

Source: CDC 1995.

Table 2.13
Module for Unmet Health Needs Survey

Availability

A) Where do you usually go for care when you have a health problem?

Accessibility

1) During the past 12 months, did you ever feel that you needed to see a doctor but could not see one?
2) During the past 12 months, did you ever go without any dental care you needed?
3) During the past 12 months, did you ever go without any eye care that you needed?
4) During the past 12 months, did you ever go without any medicines that you needed?

Need for Services/Products

5) Do you need any aids or special equipment such as canes, walkers, eyeglasses, hearing aids, or dentures that you currently do not have?
6) During the past 12 months, did you need someone to help you with your personal care, such as bathing, dressing, or grooming?
7) During the past 12 months, did you ever need help with preparing your meals because you couldn't do it yourself?
8) Do you need transportation more often than what you now have for medical appointments or getting medicines?

Source: CDC 1994.

for different age intervals as shown in Table 2.16. (See Robine and Branch [1992] for more on healthy life expectancy.)

For the total population, the life expectancy at birth in 1990 was 75.4 years; the corresponding expectancy of years of healthy life was 64. This means that the total population is expected to experience 11.4 years of less than optimal function throughout its lifetime, assuming the same mortality and health situations as experienced in 1990. This dysfunction represents the sum of the impacts of chronic conditions and injuries that occur throughout the population's lifetime, as measured by role limitation and perceived health. (Erickson, Wilson, and Shannon 1995).

Data to Build Upon 45

Table 2.14
Definitions of Role Limitation Using National Health Interview Survey Items

Not Limited
- Not limited regardless of age; this category includes unknown role performance regardless of a person's age.

Limited in Other Activities
- Limited in other activities regardless of age, or
- Limitation in activity and 65-69 years of age but able to perform activities of daily living (ADLs) and able to perform instrumental activities of daily living (IADLs).

Limited in Major Activity
- 64 years of age and younger-Limited in amount or kind of major activity, or
- 65 years and older-Major activity is considered to be ADL and IADL activities; therefore these people cannot fall in this category.

Unable to Perform Major Activity
- 64 years of age and younger-Unable to perform major activity, or
- 65 years and older-Major activity is considered to be ADL and IADL activities; therefore these people cannot fall in this category. This treatment may result in making the population less healthy than it actually might be.

Instrumental Activities of Daily Living (IADL)
- 0-17 years of age-Not applicable. Children were not asked the question about handling routine needs such as everyday household chores, doing necessary business, and shopping.
- 18-64 years of age-Unable to perform routine needs without the help of other persons and unable to perform or limited in major activity, or
- 65 years of age and older-Unable to perform routine needs without the help of other persons, or

Activities of Daily Living (ADL)
- 0-4 years of age-Not applicable. Children were not asked the question about needing help with personal care needs; therefore unable to perform their major activity is the most severe functional to which they can be assigned.
- 5-64 years of age-Unable to perform personal care needs without the help of other persons and unable to perform or limited in activity, or
- 65 years of age and older-Unable to perform personal care needs without the help of other persons.

Source: Erickson, Wilson and Shannon 1995, 47.

It is important to note that the instruments used by the U.S. Centers for Disease Control and Prevention to assess health and quality of life are self-report instruments. Thus, the conclusions drawn pertaining to the health and quality of life of the American people based on responses to those instruments are based solely on the perceptions of the people they interview. No observations

Table 2.15
Values for Health States Defined in Terms of Activity Limitation and Perceived Health Status

Activity limitation	Perceived health status				
	Excellent	Very good	Good	Fair	Poor
Not limited	1.00	0.92	0.84	0.63	0.47
Limited-other	0.87	0.79	0.72	0.52	0.38
Limited-major	0.81	0.74	0.67	0.48	0.34
Unable-major	0.68	0.62	0.55	0.38	0.25
Limited in IADL[1]	0.57	0.51	0.45	0.29	0.17
Limited in ADL[2]	0.47	0.41	0.36	0.21	0.10

[1]IADL is Instrumental activities of daily living.
[2]ADL is Activities of daily living.
Source: National Health Interview Survey, Centers for Disease Control and Prevention, National Center for Health Statistics (Erickson, Wilson and Shannon 1995, 48).

Table 2.16
Estimated Years of Healthy Life Remaining Compared to Estimated Life Years Remaining for the Total U.S. Population, 1990

Age Interval	Years Healthy Life Remaining	Life Years Remaining
0-5	64.0	75.4
5-10	60.1	75.1
10-15	55.5	71.2
15-20	50.9	66.3
20-25	46.5	61.3
25-30	42.2	56.6
30-35	37.9	51.9
35-40	33.7	47.2
40-45	29.5	42.6
45-50	25.5	38.0
50-55	21.6	33.4
55-60	18.0	29.0
60-65	14.8	24.8
65-70	11.9	20.8
70-75	9.2	17.2
75-80	6.8	13.9
80-85	4.7	10.9
85+	3.1	8.3

Source: Erickson, Wilson, and Shannon 1995, 51.

are made to determine the health and quality of life status of these people. Yet it is well known that observation instruments are often used and needed to establish the health status of individuals. In contrast, in England where QUALYs are used to assess the quality of life, the instrument on which QUALYs are based is an observation instrument (see Table 2.9 and related discussion).

The contrast does not end here. The instrument on which QUALYs are based probes two dimensions to ascertain quality of life: functional ability and distress. In this way the impact of health problems, such as disease and injury, on quality of life can be determined. Also, the effect of medical treatment on quality of life and health problems can be assessed separately. Quality of life problems and health problems can be distinguished from one another, even though they are not independent of each other. On the other hand, the quality of life instruments used by the Centers for Disease Control and Prevention do not clearly distinguish between health problems and quality of life problems. The first three questions in its quality of life Core questionnaire (Table 2.11) probe for information on health (general health, physical health, and mental health, respectively). The last question asks the respondent to relate poor health to activity limitations, which appears to be tantamount to assessing functional impairment. The QUALY instrument measures functional ability so that researchers may relate it to measures of health status statistically to study the relationship. The last question on the Core questionnaire asks the respondent to relate health status to functional ability. Isn't the relationship between health status and functional ability an important research issue?

The Quality of Life Optional BRFSS Module (Table 2.12) asks questions about limitations in activities, which I inferred meant functional limitations or impairments. The first five questions appear at first glance to be asking the respondent to relate limitations in activities or functional impairments to their health status. However, the questions seem to confuse the issue. They ask the respondent to relate health or impairment to activity limitations or impairment. For instance, question 1 asks, "Are you limited in any way in any activities because of any impairment or health problem?" The validity of responses to such confusing questions needs to be challenged. Moreover, the instruments used provide the basic data or facts upon which conclusions are based. To the extent that the instruments do not yield valid, reliable, and objective data, the conclusions based on the data cannot be accurate. Furthermore, the results of statistical analyses of any kind based on such data will be inaccurate whether the analyses yield "years of health life" or are used by sophisticated economists for cost/benefit analyses. More will be said about instruments later in the book.

Self-reports are also used to assess quality of life beyond health, including economic, political, educational, and recreational factors. Interviews were conducted in Sweden as one means of assessing the level of living or overall quality of life of the Swedish people (Erickson 1993). In another study, self-report was the only means of establishing the quality of life of Americans (Flanagan 1978). Nearly 3,000 people of various ages, races, and backgrounds, representing all regions of the United States, responded to questions in five categories (Table 2.17). However, the entire group did not respond to all questions. The results are interesting.

For approximately 431 50- and 70-year-olds responding in the study, the six areas showing the largest relation with overall quality of life were material comforts, health and personal safety, work, active recreation, learning, and creative expression. The three areas showing the smallest relation with overall quality of life are relations with relatives, having and raising children, and participation in local and national government. Those nine areas are further explained in Table 2.17. Of the many things that affect people's lives, not all of them are personally regarded as elevating the quality of life. Mature adults see the importance of being materially and personally secure and healthy, and they want to continue to learn and to do work that is worthwhile. In addition, they want to express themselves creatively in some way and to participate in active recreation. Relatives are not especially important to them anymore. Nor is having and raising children. The empty nest distress syndrome is probably overrated (Campbell 1976, 121), as is the need to participate in government and local affairs. The factors that improve the quality of life of mature adults the most are highly personal. Active personal expression and achievement seem most important, plus the health and security required to be assertive.

Other self-report instruments have been constructed to assess people's perceptions of the quality of their lives. The Life Situation Survey is such an instrument (Table 2.18). The Life Situation Survey (Chubon 1990) was carefully researched. Reliability and validity data are convincing, and it has a wide variety of uses. It can be used to compare the quality of life of different groups and to assess the impact of social policies and programs, as well as the effectiveness of clinical treatments. It is one of a very few instruments that have been found to be suitable for basic research on quality of life. It is not intended to diagnose factors that affect quality of life.

The foregoing survey of quality of life indicators and the purposes for which they are used is by no means exhaustive. Notably absent from the survey are instruments developed to assess the effectiveness of treatments on specific illnesses. Blau (1977) used quality of life indicators to evaluate the treatment of psychiatric patients. Padilla et al. (1983) developed an instrument to use with cancer patients, should the reader be interested in such highly specialized applications.

The most comprehensive list of quality of life–related instruments I have seen was compiled by Gill and Feinstein 1994 (Table 2.19). The list includes 159 instruments taken from 75 articles. Their objective was "to evaluate how well quality of life is being measured in the medical literature and to offer a new approach to the measurement" (619).

The new approach Gill and Feinstein offer and the reactions it stimulated highlight continuing controversies in quality of life assessment, especially within the medical profession. They contend that "quality of life, rather than being a mere rating of health status, is actually a uniquely personal perception, representing the way that individual patients feel about their health status or nonmedical aspects of their lives" (624). Guyatt and Cook (1994) take issue with determining quality of life solely by assessing patients' perceptions of the quality of their lives. They are "prepared to label aspects of life, such as emotional function, freedom from pain, and the ability to take care of oneself, as HRQL [health-related quality of life], without obtaining values from the individuals whose health status we are measuring" (630). The growing interest in quality of life is matched by the growing controversy over the definition of quality of life and how to observe it. If quality of life is a personal perception, as Gill and Feinstein suggest, does this mean it is fundamentally a psychological phenomenon? Does it mean that it can't be observed objectively? Are they aware that

Table 2.17
Self-Report Survey of Quality of Life in America

Components

1. Physical and material well-being
   *a. Material comforts--things like a desirable home, good food, possessions, conveniences, an increasing income, and security for the future.
   *b. Health and personal safety--to be physically fit and vigorous, to be free from anxiety and distress, and to avoid bodily harm.

2. Relations with other people
   c. Relationships with your parents, brothers, sisters, and other relatives--things like communicating, visiting, understanding, doing things, and helping and being helped by them.
   d. Having and raising children--this involves being a parent and helping, teaching, and caring for your children.
   e. Close relationship with a husband/wife/a person of the opposite sex.
   f. Close friends--sharing activities, interests, and views; being accepted, visiting, giving and receiving help, love, trust, support, guidance.

3. Social, community, and civic activities
   g. Helping and encouraging others--this includes adults or children other than relatives or close friends. These can be your own efforts or efforts as a member of some church, club, or volunteer group.
   h. Participation in activities relating to local and national government and public affairs.

4. Personal development and fulfillment
   *i. Learning, attending school, improving your understanding, or getting additional knowledge.
   j. Understanding yourself and knowing your assets and limitations, knowing what life is all about and making decisions on major life activities. For some people, this includes religious or spiritual experiences. For others, it is an attitude toward life or a philosophy.
   *k. Work in a job or at home that is interesting, rewarding, worthwhile.
   *l. Expressing yourself in a creative manner in music, art, writing, photography, practical activities, or in leisure-time activities.

5. Recreation
   m. Socializing--meeting other people, doing things with them, and giving or attending parties.
   n. Reading, listening to music, or observing sporting events or entertainment.
   *o. Participation in active recreation--such as sports, traveling and sightseeing, playing games or cards, singing, dancing, playing an instrument, acting, and other such activities.

*Factors related to quality of life.

Source: Flanagan 1978.

Table 2.18
Life Situation Survey

> **INSTRUCTIONS**: A number of statements which concern different aspects of your present life situation are listed below. Read each statement and indicate the extent to which you agree or disagree with it by checking [✔] the appropriate box in the right margin. You will note that there are six possible ratings: agree very strongly, agree strongly, agree, disagree, disagree strongly, and disagree very strongly. Do not spend too much time on each item, but try to reflect your true feelings. If you have difficulty reading the statements or marking your answers, you may have someone help you; however, only honest answers will provide useful information.

	AGREE VERY STRONGLY	AGREE STRONGLY	AGREE	DISAGREE	DISAGREE STRONGLY	DISAGREE VERY STRONGLY
1. I feel safe and secure.	☐	☐	☐	☐	☐	☐
2. My health is good.	☐	☐	☐	☐	☐	☐
3. I have too few friends who I can count on.	☐	☐	☐	☐	☐	☐
4. I like myself the way I am.	☐	☐	☐	☐	☐	☐
5. I am better off than most people in this country.	☐	☐	☐	☐	☐	☐
6. I feel constantly under pressure.	☐	☐	☐	☐	☐	☐

7. I don't eat very well. _____
8. My future is hopeless. _____
9. I am a happy person. _____
10. There are always people willing to help me when I really need it. _____
11. My income is a constant source of worry. _____
12. My sleep is restful and refreshing. _____
13. I don't get the love and affection I need. _____
14. I don't have any fun or relaxation. _____
15. Services provided by the government and other public agencies meet my needs. _____
16. I am able to go when and where I need to go. _____
17. I am satisfied with my main life role now. (for example, as a worker, student, homemaker, retiree, or patient) _____
18. There is little that I am able to enjoy in my community and surroundings. _____
19. I am exhausted well before the end of the day. _____
20. I have too little control over my life. _____

Source: Chubon 1990.

Table 2.19
Names of Quality of Life Instruments Used in the 75 Articles Reviewed

Ability to Work
Activities of Daily Living
Activity Index
Additive Daily Activities Profile Test (ADAPT)
Affective Reactions to Life
Anamnestic Comparative Self-assessment Instrument (ACSA)
Angina Pectoris Quality of Life Questionnaire (APQLQ)
Arthritis Categorical Scale
Arthritis Ladder Scale
Attitude Towards Warfarin
Body Satisfaction Scale
Bradburn Affect-Balance Scale
Cancer Instrument (ad hoc)
Cancer Rehabilitation Evaluation System (CARES)
Center for Epidemiologic Studies Depression Inventory (CES-D)
City of Hope Medical Center Quality of Life Survey
Chronic Disease Assessment Tool (CDAT) Quality of Life Scale
Chronic Disease Count
Cognitive Impairment
Colorectal Cancer Quality of Life Interview
Daily Activities
Digit Symbol Substitution Test
Disease Symptoms

Hearing Handicap Inventory for the Elderly (HHIE)
Home Parenteral Nutrition Questionnaire
HR--Quality of Life Instrument (using Multitrait-Multimethod Analysis)
Index of General Affect
Index of Overall Life Satisfaction
Index of Psychological Affect
Index of Well-being
Inflammatory Bowel Disease Symptoms Questionnaire (ISQ)
Intellectual Function (ad hoc)
Jenkins Sleep Dysfunction Scale
Kamofsky Performance Index
Katz Adjustment Scale--Relatives' Form (KAS-R)
Keitel Assessment
Kidney Disease Questionnaire
Ladder Scale (Cantrell) for Quality of Life
Lee Functional Index
Life Events
Life Satisfaction (4 domains)
Life Satisfaction (Global With Cantrell Ladder)
Life Satisfaction (Likert--7-point scale)
Life Satisfaction (10-item scale)
Life Satisfaction Index
Life Style Questionnaire
Linear Analogue Self Assessment (LASA)
Locus of Control of Behavior (LCB)
McGill Pain Questionnaire

Physical Symptoms Distress Index
Present Pain and Discomfort
Profile of Mood States (POMS)
Psychological Adjustment to Illness Scale (PAIS)
Psychological General Well-being Schedule (PGWB)
Purpose Designed Questionnaire (ad hoc)
QL-Index
Quality of Life Checklist
Quality of Life Index (QALI)
Quality of Life Questionnaire
Quality of Life Questionnaire in Severe CHF (QLQ-SHF)
Quality of Life Scale
Quality of Well Being (QWB)
Quantified Denver Scale of Communication Function (QDS)
RAND Current Health Assessment
RAND General Health Perceptions Questionnaire
Rey Auditory Verbal Learning Test
Rey-Ostereith Complex Figure Test
Rotterdam Symptom Checklist (RSCL)
Satisfaction With Life Domains Scale (SLDS)
Self Assessment Scale
Self-evaluation of Life Function (SELF)
Self Perceived Overall Quality of Life
Sentence Writing (timed)
Serial 7's

Eating Behavior (adapted from Sickness Impact Profile)
Eastern Cooperative Oncology Group (ECOG) Performance Score
Emotional Experience (adapted from RAND)
Emotional State (ad hoc)
Employment Status
EORTC GU Group's Quality of Life Form
Feelings About Present Life (Hard/Easy)
Feelings About Present Life (Tied Down/Free)
Functional Disability
Functional Living Index-Cancer (FLIC)
Functional Status (adapted from Sickness Impact Profile)
General Health Index
General Health Perceptions (GHP MOS-13)
General Health Perceptions (5-point scale)
General Symptoms
General Well-being Adjustment Scale
General Well-being Index
Geriatric Depression Scale (GDS)
Geriatric Mental State Schedule
Global Perceived Health (adapted from GHP MOS-13)
Good Days Last Week
Hand Grip Strength
Happiness
Health Assessment Questionnaire (HAQ)
Health Index
Health Satisfaction

McMaster Health Index Questionnaire (MHIQ)
McMaster-Toronto Arthritis (MACTAR) Patient Function Preference Questionnaire
Mental Health Index
Mental Status
Metastatic Breast Cancer Questionnaire
Minnesota Multiphasic Personality Inventory (MMPI)
National Institute of Mental Health Depression Questionnaire
Need for Control
Nominal Group Process Technique
Nottingham Health Profile
Other Symptoms
Overall Current Health (Adapted from RAND)
Overall Health (Global With Cantrell Ladder)
Overall Health Scale (10 cm)
Overall Life Satisfaction (OLS)
Pain Index
Pain Ladder Scale
Pain Line (10 cm)
Patient Diary
Patient Utility Measurement Scale (PUMS)
Perceived Health Questionnaire (PHQ)
Perceived Health Status
Perceived Quality of Life Scale (PQOL)
Performance Status Classification
Physical Sense of Well-being
Physical Status
Physical Symptoms (Standard Questionnaire)

Sexual Function
Sexual Symptoms Distress Index
Short Portable Mental Status Questionnaire (SPMSCQ)
Sickness Impact Profile
Side Effects and Symptoms (Hypertension)
Side Effects of Chemotherapy (ad hoc)
Sleep, Energy, and Appetite Scale (SEAS)
Social Activity
Social Difficulty Questionnaire
Social Participation Index
Social Participation (Global With Cantrell Ladder)
Standard Gamble Questionnaire
Subjectively Appraised Work Load
Subjective Rating Scale
Symptom Checklist (SCL)-90
Symptom Experience Report (SER)
Taylor Complex Figure Tests
Time Trade Off
Toronto Activities of Daily Living Questionnaire
Unfavorable External Working Conditions
Unfavorable Interpersonal Difficulties
Uniscale
Uremia Quality of Life Questionnaire (ad hoc)
Visual Analogue Scale for Global State of Well-being
Walking Test
Well-being Ill-being Clinical Observation Scale
Willingness to Pay Questionnaire
Word Recall
Work/Daily Role Well-being Scale

Source: Gill and Feinstein 1994, 622.

their definition of quality of life represents their personal viewpoint rather than a scientific conclusion based on their observations of instruments?

The research of Guyatt and Cook appears to focus on the effects of treatment on quality of life. They found in treating patients with fixed chronic airflow limitations that after using bronchodilators their patients reported that they experienced less dyspnea and fatigue during their day-to-day activities, felt more control over their illness, and had better emotional function while taking the active medication. This, they suggest, should be considered improvement in the quality of life of their patients, even though they did not meet criteria established by Gill and Feinstein. Although they dispute the criteria, they, like Gill and Feinstein, assess quality of life solely by using patient reports. Would not a measured improvement in breathing indicate improvement in quality of life? Can't patients' subjective opinions of the quality of their lives be wrong? Can professionals working to improve the quality of people's lives depend solely on people's subjective impressions of the quality of their lives?

George and Bearon (1980) offer a contrasting view of quality of life. They assess quality of life in terms of four underlying dimensions, two subjective and two objective:

Subjective Evaluations
- Life satisfaction and related measures
- Self-esteem and related measures

Objective Conditions
- General health and functional status
- Socioeconomic status

The conclusions they drew from their evaluation of quality of life instruments that assess these dimensions adds to the confusion about objective and subjective assessment. They ask, "Is objective quality of life a prerequisite for subjective assessment of quality of life? Under what conditions do people who rate low on life quality as objectively measured, nonetheless evaluate their subjective life quality in a positive manner?" (201). For additional views on the problems of conceptualizing and measuring quality of life, see Schalock (1990), Kaplan and Bush (1982), and Guadagnoli (1988). An attempt is made to clarify and constructively deal with objective and subjective assessment of quality of life in Chapter 4.

As revealed in this chapter, there are societal indicators of quality of life and similarities and differences among them. These indicators are used to ascertain the quality of life in particular geographical locations where people live for various purposes. In addition, there are personal indicators of quality of life that are used to determine health and the more general well-being of people. The status of individuals can be obtained by observers or from the individuals themselves.

I have pointed out similarities and differences of opinions between researchers and clinicians as the chapter progressed. And there was much more controversy than agreement. After surveying the research, I was disappointed to find so little consensus on quality of life indicators. The most I feel comfortable saying is that there appears to be agreement that functional disability and pain impair quality of life. Yet this seems so obvious that it is hardly worth stressing. Other conclusions that might be attempted would be less generalizable.

It was also disappointing to find that basic research does not extend much beyond the identification of quality of life indicators. If we are to improve quality of life, it will be necessary to progress from identifying quality of life indicators to identifying improvements that need to be made in order to improve quality of life. Still, we do have data to work with, and we will need to work with the data we have.

Primarily, the data we have to build upon to improve the quality of life are data on quality of life indicators with all of their shortcomings and variations. The challenge in the next chapter is to infer from and transpose these data into a knowledge base that provides a more useful foundation and guide for enhancing quality of life and defines parameters of quality of life as a field of study.

Part II

A Strategy for Improving the Quality of Life

In Part I, basic quality of life issues and challenges were explicated, and the research and development on quality of life indicators were reviewed and analyzed. Part II builds on prior research and development to describe a holistic scientific strategy for improving the quality of life.

In Chapter 3 guidelines are inferred from the existing potpourri of quality of life indicators described in Chapter 2 that identify quality of life as a field of study as well as the body of knowledge in the field. It is then shown how the guidelines can be used as a resource to promote sound decision making and meaningful research.

In Chapter 4, the limitations of the traditional scientific method are discussed, and recommendations are made for adapting scientific method to improve the quality of life. In addition, shortcomings in quality of life research are pointed out. The discussion of science in this chapter is more a presentation and analysis of basic ideas than a technical discussion.

In Chapters 5 and 6, fundamental forces of human nature are considered. To be successful a strategy designed to improve the human condition must take into account, and deal with, primary human factors. Otherwise, it is doomed to fail. On one hand, human factors present constraints on the operation of the strategy because, so to speak, "You can't make a silk purse out of a sow's ear." On the other hand, human factors can be used as resources for achieving improvement. If understood, the forces of human nature can be channeled and directed to benefit humankind. For instance, education shapes innate human intelligence into a more useful tool. It is necessary to consider two other human factors that have a primary effect on human behavior: people's motivation and their amazing intellect. If quality of life is to be improved, these two factors must be taken into account and dealt with.

We have learned a great deal about human disabilities and the constraints they impose on human functioning and how to ameliorate their impact. We have come a long way in learning how to deal with physical disabilities that impair function. We have found ways to compensate for such disabilities to help improve the quality of life of the physically handicapped. Prosthetic replacements of body parts, braces of all kinds, and compensatory mechanisms such as wheel chairs and electric scooters can compensate significantly for physical losses.

We have also learned how to deal with psychological disabilities. Now more than ever before, mentally retarded people are able to achieve their full potentials in modern educational settings designed especially for their instruction. Moreover, newer medications are more effective in relieving mental disorders such as mania and depression. We have made inroads in understanding memory loss and what to do about it. Alzheimer's disease is better understood now and may one day be relieved more effectively, if not cured.

Yet there are potent psychological forces inherent in human nature that we have not dealt with effectively in our efforts to improve the quality of life. It is well known that motivation directs behavior. Still we have not as yet applied advancements in our knowledge of human motivation to quality of life issues. It is also well known that human intelligence is responsible for our domination of the planet, yet we have not considered how it might be developed and enlisted more effectively to improve the quality of life. Instead, we become diverted by less important issues such as differences in intelligence among racial groups and whether or not the distribution of human intelligence is bell-shaped. Chapter 5 addresses motivational factors. Chapter 6 explains how human intelligence can be developed and enlisted to improve quality of life.

Finally, in Chapter 7, the parts/whole relationship among the various factors affecting quality of life are explained to clarify, relate, highlight, and summarize important elements of the proposed strategy.

Chapter 3

Quality of Life as a Field of Study

The purpose of this chapter is to derive guidelines that identify (1) quality of life as a field of study and (2) the knowledge base in the field, building on prior research. Guidelines designating quality of life as a field of study and the knowledge base in the field provide a foundation for more effectively improving the quality of life in the future. More specifically, they can serve as a heuristic foundation that supports and gives rise to more profitable scientific research and development on quality of life. Second, they can serve as a basis for making cogent decisions to improve quality of life. Thus, the guidelines can act as an aid to both researchers and practitioners.

In deriving the proposed guidelines, every effort has been made to preserve internal validity by logically inferring elements of the guidelines from the findings of prior research, much of which has been referred to in the last chapter. Once derived, the guidelines should suggest a wide variety of hypotheses that can be tested to improve the quality of life. Moreover, the database of the guidelines can be refined as the results of inquiry dictate. The derived guidelines include three dimensions: categories or areas that need to be considered when studying or making decisions about quality of life; improvements that need to be made in each area to improve quality of life; and treatments that can be used to bring about the improvements.

The first challenge was to infer from prior research more heuristic and useful quality of life categories designating areas of inquiry. Hopefully, a taxonomy composed of distinct and comprehensive categories, such as the Periodic Table and the Phylogenetic Scale, might be derived that differentiates among and covers the major factors that influence quality of life. As indicated in Chapter 2, in 1973 Smith identified six broad quality of life categories that were widely used at that time: (1) income, wealth, and employment, (2) the environment, (3) health, (4) education, (5) social disorganization (e.g., crime, alcoholism, drug addiction), and (6) alienation and political participation. Schneider (1975) concluded that "[a]ny complete study of objective quality of life would have to include at least one key variable from each of the above categories" (301). However, these categories have been found lacking. Although the six categories aid in the identification of pertinent societal quality of life variables, they are inadequate for identifying personal variables and for studying quality of life from a personal point of view. For example, Flannagan (1978), as shown in Table 2.17, found recreation to be a crucial category. So the taxonomy of six categories is not sufficiently comprehensive. Furthermore, the lines of demarcation between categories are in some cases not sufficiently distinct. "Environment" is an exceedingly broad category that to some extent subsumes the category "social disorganization," and some categories are so complex that they tend to mix apples and oranges—for example, the category "alienation and political participation."

More recently, the growing interest of the American medical profession in quality of life initiated the derivation of new quality of life categories, or domains, as they call them. On December 2–4, 1991, the Centers for Disease Control and Prevention conducted a workshop (CDC 1991). One objective of the workshop was to "achieve consensus on a working definition of health-related quality of life" (2). To meet this objective, the participants derived the quality of life domains shown in Table 3.1.

In discussing the characteristics of this model, the participants agreed that "health-related" QOL [quality of life] was generally limited to domains closer to the level of the individual, but that the societal and community domains also had important influences on individual and small group QOL, and were often included in the public's view of QOL. (3)

The domains are helpful in recognizing and identifying personal factors at the individual level. However, the participants do not indicate why they added levels or the relationship between levels and domains. Moreover, their proposal that certain factors affect quality of life at one level and not at other levels is less than realistic. For example, indicating that employment/income and recreation/leisure are quality of life factors at the societal/community level but not at the individual level is untenable, as is indicating that health and safety risks and opportunity are quality of life factors at the individual level but not at the societal/community level. Aren't there fundamental factors that profoundly influence quality of life? And aren't these factors of concern at all levels? (For the World Health Organization's and other interesting conceptualizations of quality of life, see Birren et al. [1991], Nordenfeldt [1993], Spiker [1990], and Hornquist [1982].)

After filtering through the potpourri of categories associated with quality of life previously, I was able to infer eight relatively distinctive and comprehensive

Quality of Life as a Field of Study

Table 3.1
Quality of Life Domains

INDIVIDUAL LEVEL
- Medical Condition
- Health and Safety Risks
- Functional Status
  - Physical
  - Cognitive
  - Emotional
  - Social
- Health Perceptions
- Personal Health Resources
- Opportunity
- Spirituality   (new suggested domain)
- Unmet Needs   (new suggested domain)

PERSONAL-NETWORK/HOUSEHOLD/FAMILY LEVEL   (new suggested level)
- Social Support   (new suggested domain)
- Family Functioning   (new suggested domain)

SOCIETAL/COMMUNITY LEVEL
- Peace/Freedom/Justice
- Employment/Income
- Food, Housing, Clothing
- Public Safety
- Environment
- Health and Social Services
- Transportation/Communications
- Education/Culture
- Recreation/Leisure
- Racial Equity   (new suggested domain)

Source: CDC 1991, 6.

categories that are most pertinent to quality of life from both a societal and personal perspective: government; health; work; education; remote access (via communication and transportation systems); recreation; protection; and provisions, such as food, housing, and clothing.

The categories health, work, education, and recreation are categories that have been commonly used in the past to study quality of life; the categories government, protection, provisions, and remote access have not. However, there was little question of their significance and tenability. The variables used to define each of the categories have been commonly used. For example, although remote access is a new category, communication and transportation have long been considered to be important quality of life variables. Using them to define the category remote access recognizes their common importance in providing people access to distant people, places, and things, which in many cases can affect quality of life. And although provisions is not a common category, food, shelter, and clothing are common variables used to assess quality of life. Moreover, they commonly represent material needs "to be provided for" to ensure quality of life. Protection is also not a commonly used category. Still, protection from criminals and foreign enemies is an issue often considered in assessing quality of life, as is personal security. Finally, although government is not a category that is used commonly to assess quality of life, variables controlled by government commonly are—for example, freedom and political participation. The eight categories identify distinctive and relatively comprehensive generic factors that influence quality of life. In addition, variables in each category have been associated with quality of life in the past. These issues will become clearer as the categories are discussed further.

The second challenge was to identify some improvements in each of the eight areas that might be made to improve the quality of life. Other improvements can be added as warranted. The point is that if we are to improve the quality of life, we need to move beyond identifying indicators of quality of life and identify specific improvements that need to be made. It is more likely that quality of life will be improved by finding treatments to effect needed improvements than continuing to expand the vast list of quality of life indicators. I found in inspecting the many quality of life indicators that have been identified that the desired direction for improvement was often implied. For example, in inspecting the scales measuring functional ability, it was quite clear that movement from less to more mobility constitutes improvement. And movement from less to greater work satisfaction constitutes improvement. So in many cases I had little difficulty inferring needed improvements from quality of life indicators.

The third challenge was to derive treatments that can be used to bring about the improvements identified. Most often, treatments are interventions or manipulations devised to achieve particular outcomes. However, sometimes the prescribed treatment can be to do nothing. Such is the case when a problem can be expected to dissipate with the passage of time. Treatments were derived in two ways. Sometimes in studies of quality of life indicators, appropriate treatments were mentioned or alluded to. In addition, it is often possible to infer from the description of an improvement the characteristics of a treatment for effecting the improvement. Most often, general treatment characteristics can be inferred rather than specifics of a treatment. Hopefully, through research and development, the effectiveness of treatments can be increased, thereby improving the quality of life.

Guidelines for improving the quality of life were derived by first considering improvements that can be made and then deriving treatments to effect the improvement in each of the eight areas mentioned. Since improving the quality of life in societies is a somewhat different challenge than improving the lives of

Quality of Life as a Field of Study 65

individuals, separate guidelines were derived for each. The guidelines are logically derived from prior research.

GUIDELINES FOR IMPROVING THE QUALITY OF LIFE IN SOCIETIES

The guidelines for improving the quality of life in a society specify particular improvements that are frequently needed to make a society a better place to live. Indicators of improvement are normative to reflect general living conditions in a society. For example, in the category education, increasing the literacy rate in a society might be an index of improved education. The treatments described are prescriptions for achieving the improvements specified. For example, increasing the number of teachers per capita might be a proposed treatment to reduce the illiteracy rate. An effort was made to identify basic improvements that are often needed in each of the eight categories and to infer reasonable treatments to achieve the improvements.

Government

Since government has not been designated as a quality of life category in prior taxonomies of major quality of life factors, some discussion is needed to introduce it. Government is an overriding factor in a society because the laws that are enacted or proclaimed and enforced determine to a great extent other factors such as education, protection, and health care. Free choice, or civil liberty, is used as an indicator of quality of life in nations (see Chapter 2). Although freedom may be a precious commodity, many in free countries regard it as sacrosanct and are intolerant of constraints. If freedom is to be promoted effectively, it must be promoted realistically.

Liberating the oppressed may deserve to be high on the human agenda. However, it is important to realize in trying to redress the transgression that there can be too much freedom. An increase in freedom is quite often an improvement—but within limits. The goal, and the challenge, is to achieve a delicate balance between laws that ensures personal autonomy and laws that promote the common good. They often conflict. When the scale tips too far in favor of the common good, people may have to sacrifice fulfillment of their personal aspirations and find whatever satisfaction they can by pursuing imposed societal objectives. This tends to be the pall in totalitarian, socialist, and communist societies. On the other hand, neglecting the common good to increase personal autonomy engenders overpermissiveness, which undermines the foundation of society and the security of its people. Force and fraud must be minimized to maintain government, commerce, and personal freedom. Laws must be enforced to protect people from being coerced by physical force and the lies of others for government, commerce, personal freedom, and domestic tranquility to be tenable. Indicators that have been used to assess civil liberties in societies are freedom of speech, press, assembly, and religion. Indicators used to assess protection against force and fraud are violence and bunco crime rates.

Among the various privileges of freedom, free enterprise is lauded for raising the standard of living in societies. Free enterprise does increase productivity and abundance, but it increases exploitation and turmoil, too. In contrast, socialism and communism provide welfare for the needy and tend to subdue violence, because when more people work for the common good, there seems to be less upheaval. But socialism and communism do not promote productivity, as free enterprise does.

There appears to be confusion that needs to be cleared up. Democracy and totalitarianism are mutually exclusive, and free enterprise and communism are mutually exclusive. Democracy and communism need not be mutually exclusive. It happens that democracies enact communistic laws. Mandates in Karl Marx's *Communist Manifesto*, such as income tax, have been democratically voted into law in the United States to share the wealth of the haves with the have-nots. And totalitarian nations, such as Red China, are allowing free enterprise.

Political opportunity was another attribute used to assess quality of life in nations. It deserves attention because it is often overlooked, and there can be little freedom without it. Political opportunity is necessary to maintain a democratic government. To maintain a democracy, there must be political opportunity to vote, to conduct honest elections, and to run for government office and for minority groups to influence people and government. These are four indicators used to assess political opportunity in nations.

Once needed improvements in government are specified, the treatments required to achieve them emerge. General guidelines for devising treatments include enacting and enforcing laws that ensure political opportunity and free choice without unduly sacrificing the common good.

Health

Health is a second area in which improvement is most often needed and sought. Four variables are commonly measured to assess health in a society: longevity, absence of controllable ailments, mental stability, and functional ability. Longevity pertains to the quantity of life. The other three factors pertain to how healthy people are while they are still alive. Incidence of controllable ailments indicates whether people in a society are suffering from ailments that can be prevented and treated. Presence of such ailments, for example, cholera, polio, or deformed bones, represents social deprivation. Functional ability of the citizenry indicates the fitness of a society, provided that functional ability extends beyond the absence of disability. Functional ability must include strength, endurance, coordination, and mobility. Mental stability is included because an index of mental health is needed to complete an assessment of health. Table 3.2 shows the four variables that need to be considered to assess health in a society as well as measures of the variables. Other measures can be used as well.

Other indicators used to assess health in societies include mortality rates, number of accidents per capita, life expectancy, and drug addiction rate.

**Table 3.2
Health**

Variables	Measures
Longevity	Average life expectancy
Mental Stability	Suicide rate
Functional Ability	Strength, endurance, coordination, and mobility of the average citizen
Incidents of Controllable Ailments	Number of treatable and preventable ailments per capita

Source: Author.

Health care treatments needed to improve health in a society include increasing the number of health care professionals. The right mix of medical specialists and general practitioners, plus nurses, physical therapists, occupational therapists, and vocational counselors, needs to be attained. In addition, hospital and nursing care beds need to be increased and located to serve more people. There is a great lack in delivering available medical care to people, especially those who live in city slums and rural areas. Health care professionals in all fields must be provided advanced, modern facilities, equipment, and medications. Finally, it is society's obligation to control environmental pollution that jeopardizes health.

Work

Improvement in work centers around increasing the purchasing power of workers, rather than increasing their wages. The most important issue is what people in a society can buy on the average for a unit of time at work—for example, the amount of food that can be purchased for an hour's work. Other indicators that might be used are not as central in assessing a society's responsibility in the area of work. People work for employers or are self-employed so that they can purchase what they need and want for themselves and their loved ones. Societies can make it possible for their citizens to reap more for their work. Another improvement that societies can effect is reducing the unemployment rate. People must be employed to have any purchasing power at all. Other indicators used to assess working conditions in a society are gross national product, fatal work accidents, percentage of children in the labor force, and average workweek per capita income.

Suggested treatments for improving purchasing power are to increase high-tech manufacturing and services and the number of businesses in a society. High-tech manufacturing and services create higher-paying jobs. Increasing the number of businesses in a society increases both the number of employers and the number of employment opportunities. This, in turn, decreases the unemployment rate. Unemployed people have no purchasing power and become societal wards. In addition, in modern societies that use money as a medium of exchange, inflation must be kept under control. Although in most societies inflation is a fact of life, decreasing the rate of inflation reduces the erosion of purchasing power.

Education

Education is the means by which societies perpetuate themselves. It can be improved on, on two dimensions. First, the percentage of the population educated can be improved. This requires that children start schooling at a young age and stay in school until they receive a certificate attesting that they are prepared for their social responsibility and gainful employment. Both the illiteracy rate and the dropout rate are serious problems in the United States. Table 2.4 shows that the United States is not included among the top ten nations with respect to literacy. And the dropout rate in the United States reaches as high as 50 percent in some inner cities (Friedman 1993). Second, the average level of education of the citizenry needs to be elevated in many societies. Many occupations crucial to the quality of life in society require advanced education, for example, the training of doctors, engineers, and research scientists. Students should not leave school until they are literate and employable. Higher-paying jobs critical to modern society require advanced education. Indicators used to assess education in a society include literacy rate, student/teacher ratio, and number of college graduates per capita.

Treatments required to attain the education improvements described include increasing the number of teachers employed per capita, which in turn should decrease the student/teacher ratio. In some societies the elderly are recruited to teach at relatively little expense. In addition, the quality of life in a society will improve when a higher percentage of teachers can teach high-tech skills such as doctoring, engineering, and research skills. Finally, increasing the number of libraries and access to them is an important factor in improving education. Libraries include data banks as well as books and periodicals on the shelf. Libraries serve as teachers' aids as well as allowing people to educate themselves.

Remote Access

A major difference between primitive and modern societies is that people living in modern societies have much greater access to distant people, places,

Quality of Life as a Field of Study

and things through better communication and transportation facilities. Improvement in communication in a society is manifested by an increase in the average number of messages that are sent and received by people who are not in face-to-face contact with each other. In primitive societies, remote contact is usually limited to smoke or drum signals, which do not reach beyond visible distance or earshot. Modern means of communication span the globe.

Treatments to improve communication involve increasing the availability of communication devices to the citizenry, such as phones, computers, fax machines, postal services, and newspapers.

Transportation is also vital to improve access to distant people, places, and things. Unlike communication, transportation allows people to come into face-to-face contact with people not within walking distance and to send merchandise to, and receive products from, far-off places. Two indicators of improved transportation in a society are the average distance traveled per person a year and the amount of freight received from and shipped to places outside the limits of the society.

Treatments required to effect the improvements include increasing the number and types of vehicles available to the citizenry, the shipping facilities, and public transportation available to people, plus reducing the cost of travel and shipping merchandise.

Remote access is often studied by assessing indicators such as the number of phones per capita, number of vehicles owned per capita, length of roads per square kilometer of territory, radio receivers per 1,000 people, average cost of transportation for a family of four, and number of daily newspapers. Although these indicators are studied, remote access has, to my knowledge, never been identified and studied as a major factor influencing quality of life, and it may well be an important factor.

Recreation

Engaging in games, sports, arts, and crafts, both as participant and as spectator, brings enjoyment to people and provides respite from work and tense social relations. Although favorite pastimes vary from society to society, people always make time for recreation. Improvement in recreation is manifested as an increase in the percentage of the population engaged in recreational activities.

Treatments to improve involvement in recreation entail increasing the number of recreational facilities and opportunities available to the public, such as parks, television and radio stations, tournaments, museums, and art galleries.

Protection

A major function of a society is to protect its citizens from criminals within the society and from foreign enemies. People who don't feel safe have difficulty working, learning, and maintaining amiable social relations. A decrease in the crime rate as well as in foreign incursions reflects improved protection.

Treatments to improve protection entail increasing the number of law enforcement officers per capita, adding jails and crime prevention programs as needed, and updating the military equipment of the armed forces.

Indicators of protection used in research include soldiers per 1,000 civilians and crimes per capita. Crimes per capita indicates a possible improvement needed. Increasing the size of the military can be regarded as a treatment used to protect people against foreigners. However, as an indicator of a possible improvement needed, too many soldiers is often regarded negatively as indicative of a repressive police state. Armed protectors have been oppressive to the people they are empowered to protect, and many people fear and try to guard against the abridgement of political freedom that can result from overly intrusive protection. So armed protection and personal freedom can be antagonistic.

Provisions

For a society to thrive, it is necessary that its citizens be able to avail themselves of the products, services, and housing they need and want. Improvements in provisions include increases in the number and variety of products, services, and residences available to the citizenry. Reducing the average number of people per room is an index of improved housing.

Treatments to improve provisions include increasing productivity and distribution of modern products and services. Housing can be improved by building more and larger residences. Adding thermostatically controlled air conditioning and/or heating as needed is another residential improvement. Reducing the cost of provisions will enable people to purchase more of them.

Other indicators used to assess adequacy of provisions in studies include average size of residence, percentage of substandard dwellings, energy consumption per capita, daily calorie consumption, and discretionary spending (see Table 2.4).

Table 3.3 summarizes the guidelines for improving quality of life in societies. Other improvements may be identified in each category as well as treatments for effecting the improvements.

The preceding guidelines for improving the quality of life in societies focus on making nations, states, cities, and communities better places to live. The following guidelines focus on improving the quality of life of individuals.

GUIDELINES FOR IMPROVING THE QUALITY OF LIFE OF INDIVIDUALS

The improvements identified in each of the eight areas are important to individuals, whatever the society they live in may be like and allowing that the society people live in affects their personal lives. Looking at quality of life from a personal perspective is quite different from looking at quality of life from a societal perspective. The quality of life of individuals can be quite good in a

Quality of Life as a Field of Study

society in which the quality of life is poor on the average. Normative information such as averages, percentages, and rates have limited application and can be spurious. For instance, it might be said that on the average a person who has one foot on a hot stove and the other in a freezer is on the average comfortable.

Government

To some people, government is "Big Brother" whose primary purpose is to take care of them. To others, it is a meddlesome, inefficient bureaucracy imposing its will on them and thwarting their ambitions. Whatever the nature of their government and whatever their personal sentiments toward government may be, people will fare much better in any society if they have knowledge of their legal rights and limitations and how to obtain their entitlements. They don't need to be experts on the law, but they do need to know enough to take advantage of their legal entitlements and to avoid being arrested. In addition, they need to be able to distinguish between the laws of the land and the expectations of social groups that they may accept or reject as they choose. Moreover, people need to learn how to acquire knowledge about the law and advocacy when they find themselves in unfamiliar predicaments.

No previous quality of life studies could be found in which people's knowledge of their legal limitations and legal entitlements and how to obtain entitlements were researched. There is reason to believe that people will feel more secure and content when they know what to expect and what they can and cannot control, as will be explained later in the book. Some studies assess people's participation in voting, campaigning, and lobbying but not much else in the realm of government.

Treatments to increase people's knowledge of the law include teaching them the legal limits of free choice and how to take advantage of legal entitlements. People should also be informed of the resources they can use to find out more about the law as need be. Basic education in the law can begin early in school. Instruction on resources would include how to access publications that explain the law in lay terms, how to find advocacy groups that help people acquire their legal entitlements, and when and how to find lawyers that can be of help.

Health

From a personal point of view, health involves more than the absence of injury, disease, and mental instability. Health has a positive dimension as well. People can be hale and hardy and be content with their lives. Overcoming disability involves recovering from mental and physical ailments. When people are not suffering from disability, they can work to improve their strength, endurance, coordination, mobility, and mental acuity. In addition to improving their mental and physical abilities, they can work on improving their interpersonal relations

Table 3.3
Guidelines for Improving the Quality of Life in a Society

Areas	Improvements	Treatments
Government	Increase free choice and political opportunity.	Enact and enforce laws that ensure political opportunity and free choice.
Health	On the average, increase longevity, functional ability, mental stability. Decrease controllable disease.	Increase the number of health care professionals and hospital beds. Provide advanced health care, facilities, and medication. Control pollution.
Work	Increase average purchasing power.	Increase high-tech manufacturing and services and the number of businesses. Decrease rate of inflation.
Education	Increase percentage of people educated and the average level of education.	Increase the number of teachers employed, their ability to teach high-tech skills, and the number of libraries.

Remote Access	Increase the average number of messages and shipments sent to and received from distant places and the distance traveled per person.	Provide a greater number of communication devices, such as phones, fax machines, postal services, computers, newspapers. Increase the number and variety of vehicles. Decrease the cost of travel.
Recreation	Increase the rate of recreational involvement.	Increase the number of recreational facilities, such as parks, television and radio stations, and tournaments.
Protection	Decrease the crime rate and foreign incursion.	Increase the number of law enforcement officers, jails, and crime prevention programs. Upgrade the equipment of the armed forces.
Provisions	Increase the number and variety of products, services, and residences. Reduce the average number of people per room.	Increase production and distribution of products, services, and housing.

Source: Author.

and their contentment with life. Physical health, mental health, and healthful social relations are all important to personal health. They are not independent of each other. All three need to be attended to.

A great deal of research has been devoted to the study of health, and many instruments have been developed that assess variations between extreme disability and no disability (see Chapter 2). Although there are measures of degrees of strength, endurance, coordination, and mobility, they tend to be developed more by professionals in athletics and physical education than by health care professionals. The full range of functional abilities needs to be assessed in order to determine how healthy people are and to encourage them to work to become vigorous, hardy, and buoyant.

Treatments to improve personal health are primarily educational. People need to be taught how to acquire information on the effectiveness of treatments typically prescribed for particular ailments and their side effects as well as how to acquire the available health care they need. To become vigorous, hale, and hardy, they need to be taught healthful diet and exercise regimens. Teaching can be a part of formal schooling. It should include not only basic health care information but also guidance on how to access information from periodicals, books, and data banks to keep abreast with the times.

Work

Whatever the purchasing power of money may be in a society, an increase in earnings will improve the quality of life of individuals. So a central index of improved employment is an increase in earnings. Another important index of improved work is an increase in work satisfaction. Since people work a good part of their days, work satisfaction contributes substantially to life satisfaction. When people dislike their jobs, they find it difficult to get out of bed in the morning, regardless of how much they earn. Further, people who work at home without pay must derive satisfaction from it, directly or indirectly, or they have little reason to do it. An indirect measure of work satisfaction that is worth investigating is accumulated wealth in the form of savings and investment. It might be hypothesized that people who save more money are more satisfied with their work lives.

Income is quite often used to assess quality of life in studies—but not necessarily in the category of work. It is often related to material well-being (see Table 2.17). Work satisfaction is routinely studied in business and industry, and instruments are available to measure it. Still, no quality of life studies could be found that assessed work satisfaction as an attribute of personal well-being. Other indicators of favorable working conditions used in studies include time off, physical demands of work, perks, stress, and safety.

Treatments, again, include education. People need to be taught how to make work choices among self-employment, government employment, and employment in the private sector. In considering job possibilities, they need to be taught how to weigh working conditions, perks, and opportunities. Most important, they must be taught to consider which positions suit them, whether they would be content doing the work, and qualifications for the job. There is also a need to learn how to plan for career changes and advancements. Increased longevity and changes in job market demands make it imperative that people prepare to make career changes. They must be taught how to advance in a particular career and how to change careers. Education is also important to prepare people for work on the job or at home.

Education

On a personal level, education is the means by which individuals prepare to fulfill their aspirations and succeed in society. Basically, any increase in knowledge and skill constitutes an improvement in education. Knowledge seems to be sufficiently emphasized in most Western societies, judging from the amount of memorizing students are required to do. Skill development is frequently neglected. Ultimately, the success people have depends on their skills in controlling themselves and their environments.

Indicators of educational achievement in prior research include years of education and level of education reached. Studies that attempted to find out from people the extent to which education is important to the quality of their lives inquired about the importance of learning, attending school, improving their understanding, and acquiring additional knowledge.

The treatment for instilling knowledge and skills is instruction on how to fulfill personal aspirations within necessary constraints and how to make social contributions. Since education is the means of perpetuating and indoctrinating youth into society, students must be taught the language and laws of their society so that they can be socially productive, law-abiding citizens. To succeed both personally and socially, the teaching of skills must be emphasized. It takes more than knowledge to achieve success. People succeed primarily because of what they can do—their skills. And the perfection of skills depends on predictive ability. People must be able to predict the consequences of their actions to be able to achieve any objective. People who can't are incompetent, insane, or both. More on predictive ability will be given later.

Remote Access

The more access people have to other people, places, and things both near and far from them, the better their chances of getting what they want. Access to nearby things is available in primitive societies. One of the benefits of modern societies is access to more remote people, places, and things. So remote access appears to be a hallmark of human progress and a benefit people seek.

Improved remote access is evidenced by an increase in individuals' ability to communicate with people whom they cannot contact face to face. Improved remote access is also evidenced by more effective transportation, that is, by an increase in the ability to travel and to send and receive merchandise to and from distant places.

Although the category of remote access does not appear in prior quality of life research, indicators such as owning a telephone or car frequently do appear. Treatments for improving remote access include instruction on communication and transportation, that is, how to use languages and how to acquire and use available communication devices such as phones, fax machines, and computer networks, as well as how to acquire and use vehicles and available transportation systems.

Recreation

Recreation is exceedingly important to people. Many would rather do without toilets, bathtubs, and household appliances than to do without a television set. And as shown in Chapter 2, mature adults want very much to be actively involved in sports, arts, and crafts. Improved recreation is manifested by increased enjoyment from activities. Recreation is personal. People choose it voluntarily to please themselves. In contrast, the work people do most often is done to please someone else, for instance, the boss or customers.

Recreation is almost always included in quality of life studies of individuals. Typically, studies probe preferences for socializing, such as joining clubs and attending parties; passive recreation, such as listening to music or observing sporting events or entertainment; and active participation in sports, games, or traveling.

Treatments include instruction on available amusements, sports, games, clubs, arts, and crafts and how to partake of them.

Protection

On a personal basis, protection is appreciated because it enables people to be and feel secure. When people fear for their lives and property, they are preoccupied with self-preservation. They regard their environments as hostile they are haunted by the transgressions they anticipate, and their functioning is inhibited. Most families naturally try to protect one another. Improvement in personal protection in childhood pertains to improved parenting and protection in school. In adulthood, it pertains to improved access to trustworthy neighbors and police and fire departments and improved insurance coverage.

Protection, as a category, has not been studied in prior research to determine the quality of life of individuals. However, people are sometimes asked about the importance of safety or security to them or whether they feel anxious, distressed, or threatened.

Treatments that promote safety include instruction on how to protect life and property and how to utilize such devices as burglar alarms, neighborhood watches, protection agencies, fire and police departments, and insurance policies.

Provisions

People spend a good portion of their money for food, shelter, and clothing. An increase in their knowledge of available food, clothing, and housing and how to acquire them serves to improve their chances of obtaining the provisions they want and need. These provisions, once thought of as basic necessities, are sought now in many modern societies as status symbols. Buying chic clothes, living in the right neighborhood, and partaking of vintage wines and gourmet food are in modern societies associated with financial success and social status.

The category of provisions does not appear in studies of the quality of life of individuals. However, research inquiries are made about housing and nutrition. Food, housing, and clothing seem to be addressed more often in studies of societies—for example, average calories consumed per capita or average size of residences.

Treatments to improve the provisions people acquire include instruction on cost, quality, and availability of products, services, and housing. Table 3.4 summarizes the guidelines for improving the quality of life of individuals. Instruction is an overriding factor enabling individuals to improve the quality of their lives. The preeminence of instruction might be surprising at first. However, after serious consideration, it should not be. Although the benefits societies bestow upon their citizens vary substantially from society to society, the benefits acquired by any individual in a given society are dependent on the instruction that person receives from parents and other teachers. Most of the adaptive responses in any adult's repertoire have been learned. Moreover, most preventative treatments utilize instruction as a primary intervention—for example, the prevention of drug abuse and coronary disease. And instruction is often used to influence people to discontinue harmful practices such as cigarette smoking.

For those interested in a definition of quality of life, we can at this time consider categorical or dictionary-type definitions. As the reader may know, in deriving categorical definitions, a category is given a word label, and attributes for inclusion in the category are enumerated so that members or examples of the category can be distinguished from nonexamples. The category we are concerned with is labeled "quality of life," and many attributes of quality of life have been discussed. How does one select attributes to derive a definition? Would not any definition be arbitrary? It seems that it would. To complicate matters further, we have considered two types of attributes: societal and personal attributes. Does this mean that two definitions of quality of life are needed? Perhaps it does. Consider a personal definition first. One attribute may be sufficient for a personal definition of quality of life. For example, it might be said that from a personal point of view the quality of a person's life depends primarily on his or her "ability to pursue personal aspirations." In contrast, it

Table 3.4
Guidelines for Improving the Quality of Life of Individuals

Areas	Improvement	Treatments
Government	Increase in people's knowledge of their legal rights and limitations.	Instruction on the legal limits of free choice and how to take advantage of legal entitlements.
Health	1) Recovery from mental and physical ailments; 2) improvement in strength, endurance, coordination, mobility, interpersonal relations, and mental stability.	Instruction on 1) outcomes and side effects of therapeutic treatments and how to obtain appropriate treatment; 2) diet and exercise regimens; 3) interpersonal skills; 4) stress prevention.
Work	Increase in earnings and work satisfaction.	Instruction on how to make work choices and career advancements.
Education	Increase in learning knowledge and skills.	Instruction on how to 1) fulfill aspirations within necessary constraints; 2) make social contributions; 3) improve predictive ability.

Remote Access	Increase in the ability to communicate from afar, to travel, and to transport merchandise.	Instruction on how to 1) use languages; 2) obtain and use communication devices; 3) acquire and use vehicles.
Recreation	Increase in enjoyment.	Instruction on available amusements, sports, games, clubs, arts, and crafts.
Protection	Increase in security.	Instruction on how to protect life and property.
Provisions	Increase in knowledge of available products, services, and housing and how to acquire them.	Instruction on the cost, quality and availability of products, services, and housing.

Note: In general, treatments involve instruction.
Source: Author.

might be said that a society needs to possess many attributes to be a good place to live. Thus, quality of life in society might be defined by enumerating any or all of the following attributes.

Factors	Attributes
Government	Opportunity for (1) free choice and (2) political participation
Health	Control of (1) disease, (2) functional ability, and (3), pain
Work	Opportunity to increase (1) purchasing power and (2) work satisfaction
Education	Opportunity to acquire (1) knowledge and (2) skills
Remote access	Opportunity for remote (1) transportation and (2) communication
Recreation	Opportunity to participate in (1) games (2) hobbies (3) arts and (4) entertainment
Protection	Control of (1) crime and (2) foreign incursion
Provision	Opportunity to acquire (1) food, (2) clothing, and (3) housing

Now, it should be understood that scientific research can proceed without general agreement on the definition of terms. Researchers simply provide operational definitions of the terms they are using and proceed. Typically, this amounts to stipulating the procedures they are using in manipulating and observing variables and in establishing relationships among variables for the purpose of enabling others to understand and replicate their research. More will be said about operational definitions in the next chapter.

Scientific definitions tend to change with the acquisition of knowledge over time. Lasting generalizable definitions usually are associated with more advanced scientific disciplines such as physics and chemistry. Quality of life as a discipline is still embryonic. For the time being, we need to be content with working definitions of quality of life, which can be derived from the guidelines presented.

Presentations in the last chapter make it quite clear that variables in the guidelines (Tables 3.3 and 3.4) have been studied in the past. However, the proposed format is new and has not been systematically applied. A comprehensive, synchronized approach is advocated in which research is conducted in all eight areas to determine treatments that will bring about particular improvements in quality of life. Studying treatment → improvement links is a productive approach to improving the quality of life much more so than studying improvements or treatments independently. Endless amounts of time can be spent defining and constructing measures of quality of life and studying quality of life indicators without ever improving quality of life. Moreover, although identifying improvements that need to be achieved is necessary to improve quality of life, it is not sufficient. Treatments that will bring about the improvements must be determined as well.

Furthermore, it is fruitless to advocate treatments without being able to specify the particular improvement the treatment can be expected to achieve. Still, many available treatments claim general benefits, with no proof that they affect any general or specific improvements. For instance, spiritual cults claim that abiding by their doctrines elevates overall quality of life. But like all panaceas, they claim more than they can prove they can deliver.

Over many years, researchers working to construct a conceptual framework to improve quality of life have attempted to identify (1) comprehensive and distinct categories representing major factors that affect quality of life and (2) significant quality of life indicators within each category. The proposed guidelines summarized in Tables 3.3 and 3.4 represent an extension of their work. Perhaps a practical application will illustrate the value of such guidelines in general and the proposed guidelines in particular. Let us consider the work of the Centers for Disease Control and Prevention as a case in point. The following application of the proposed guidelines brings to light shortcomings in the CDC's approach to quality of life. It is not meant to detract from the very valuable work it does in controlling and preventing disease.

First, the CDC's specifications of domains or categories are not clear. Definitions of the domains are not specified, and the addition of levels to domains confuses the issue. The definition of domains could be clarified by enumerating quality of life indicators within each domain.

Second, the CDC's conclusion that "health-related quality of life was generally limited to domains closer to the individual" (CDC 1991, 3) is questionable. The CDC is a government agency funded by the federal government. The data it is concerned with are largely normative or societal data. It may be true that the most fundamental source of health data pertains to individuals, but the CDC pursues societal health objectives. Additionally, the CDC's ability to provide effective health care suffers from overly restricting its focus. Although the mission of the CDC is to improve health, the means of improving health often require more than medical treatment. Other factors may need to be dealt with as well. Education may be needed to prevent disease. Government subsidy may be needed to provide funds for treatment. Recreational activities may be prescribed to ensure sufficient exercise or to improve mental health. A work change may be needed because of disability limitations. Provisions may need to be prescribed if, for example, a change in diet is required. And protection might be an issue for frail or disabled people. The objective of the CDC might be limited to improving health, but treatments required to achieve the objective may pertain to governmental, work, educational, recreational, and other factors specified in the guidelines and discussed. It is crucial to distinguish between means and ends.

Third, the categories the CDC specifies within the individual level are questionable:

Individual Level
    Medical condition

Health and safety risks
Functional status
- Physical
- Cognitive
- Emotional
- Social

Health perceptions
Personal health resources
Opportunity
Spirituality
Unmet needs

The distinctness and comprehensiveness of these categories are suspect. They compromise comprehensiveness when they exclude other levels of inquiry. Additionally, other important categories on the individual level are missing. Are there not other categories that represent major health-related factors that affect individuals' quality of life, for instance, their knowledge of hygiene, illness symptoms, and how to obtain medical treatment? As for distinctness, the distinction between medical condition and health perception is not clear.

Fourth, the CDC appears to confuse the distinction between domains or categories and indicators. For instance, it regards functional status as a domain, rather than as a quality of life indicator. The review of functional status instruments in the last chapter shows that functional status is an indicator that is observed routinely. It is no small matter to confuse categories and indicators. Indicators are variables that can be observed. Categories exist by virtue of definition and are defined by enumerating the indicators of categorical membership. A category cannot be observed. Only examples or instances of categories can be observed. For example, *cancer* is a categorical term that is defined as a syndrome of indicators. Cancer cannot be observed directly; only the syndrome of indicators that define cancer can be observed. A case of cancer is identified when the indicators of cancer are observed in a person. Since science depends on observation, it is crucial that we understand what can and cannot be observed. And since indicators, not categories, can be observed, the relationship between categories and indicators must be clarified before observation instruments are selected or developed to observe indicators.

Fifth, inspection of the instruments the CDC is using to assess quality of life, the Core, Optional and Unmet Needs instruments, raises concerns about the data being collected. Data on perceptions of unmet needs are obtained from responses to the unmet needs questionnaire (Table 2.13). Some data on perception of functional status are obtained from responses to the optional questionnaire (Table 2.12). And primarily data on health perceptions are obtained from responses to the core questionnaire (Table 2.11). There appear to be two major shortcom-

Quality of Life as a Field of Study   83

ings in the data being collected. First, the responses to all three questionnaires are perceptions, which are not necessarily indicative of actual conditions. Are they not using scientific tests to obtain data on quality of life? Is the CDC implicitly endorsing the position that quality of life is solely a matter of personal perception that exists only in the minds of people? Is it implying that quality of life cannot be observed by scientific instrumentation? This would indeed be a biased and very limiting approach to quality of life, implying that quality of life cannot be studied scientifically, as disease can.

Second, the questionnaires elicit data in a limited number of the domains the CDC has designated as important to health-related quality of life. How is the CDC obtaining data in the domains of medical condition, health and safety risks, personal health resources, opportunity, and spirituality? In addition, although functional status has been designated as a separate domain, questions were raised in the last chapter about the questionnaires' ability to distinguish between health status and functional status.

Sixth, according to the CDC, it wants to move beyond assessing quality of life and is committed to improving the quality of life in America:

In 1989 the mission statement of the CDC was revised to read as follows: To improve the quality of life for all Americans by preventing unnecessary disease, disability and premature death and by providing healthy lifestyles. This revision constituted a significant change from past mission statements in that it identified health-related quality of life improvements as a primary goal of the agency. (CDC 1991, 1)

To achieve this mission, the CDC needs to specify specific improvements it wants to achieve. This entails more than stating general mission statements and goals. Such statements may serve to reflect general intentions for conversational and public relations purposes, but little else. If the CDC wants to improve quality of life, it needs to specify particular improvements it proposes to achieve in operational terms. The definition of *improvement* will be discussed in detail later. For now, it is sufficient to say that an improvement is progress from an existing state to a more desired state. Those who want to improve quality of life need to designate the progress from particular existing states (starting points) to particular desired states (goals) they want to achieve.

A review of some of the CDC's findings presented in the last chapter brings to light the need (1) to specify particular improvements the CDC is interested in pursuing, (2) to obtain observation instruments for assessing the extent to which the improvements are needed and (3) to develop treatments to bring about needed improvements. To illustrate important problems, let us consider just one question on the core quality of life questionnaire. The first question on the questionnaire—"Would you say that your health is excellent, very good, fair, or poor?"—asks for respondents' perceptions of their health. Based on responses to this question, it was concluded that Americans' perceptions of their health vary from state to state, and relative perceptions among the states were described. What are people supposed to conclude from this report? Are people's perceptions supposed to be construed as actual conditions? Is it being suggested that if people living in a low-ranking state want to improve the quality of their lives, they should move to a state in which people's perceptions of their health ranks high? What particular improvements that the CDC wants to achieve are these findings relevant to?

Another report of responses to this question indicates that "86.6% of respondents reported good to excellent self-rated health" (CDC 1995, 196). The *New York Times* ("87% of Americans Say They Feel Healthy," 1995) reports a doctor at the CDC as saying, "This is good news, in sharp contrast to the usual messages we hear of gloom and doom about health. People in general think their health is pretty good" (17). How is this finding related to improvements the CDC wants to achieve? Are the findings supposed to indicate that there is little room for improvement?

Since the improvements the CDC wants to achieve are not clear, and there was no indication that particular questions being asked are related to the assessment of particular improvements, it is impossible to know how responses to the questions on the questionnaire are to be interpreted. The responses to question 1 on the core questionnaire are offered as an illustration; there are problems with other questions as well. Good professional advice given to those constructing questionnaires is, If you don't know how you will interpret the answer to a question, don't include the question. In the present context, this means that the improvements to be achieved need to be specified before instruments are selected or developed to assess indicators of the improvements. In this way, data collected from the administration of the instruments can be interpreted as indicating the extent to which the improvements have been achieved.

The preceding example illustrates the importance of developing a cogent conceptual framework to guide inquiry and the interpretation of the results of inquiry. At this point, we are ready to infer, from the derived guidelines, quality of life as a field of study or discipline. The field of study is circumscribed and illustrated in Table 3.5. The major areas of study within the discipline are government, health, work, education, remote access, recreation, protection, and provisions. Quality of life is analyzed in all eight areas from societal and personal perspectives. In this way, relationships between societies and individuals can be more clearly seen. At the same time, differences between societal and personal influences become more apparent.

The basic unit of analysis, and the focus of inquiry in the field, is the treatment → improvement relationship, in which improvement is the effect to be achieved (dependent variable) and treatment is the causal agent (independent variable) used to achieve the effect. This basic unit of analysis is studied in all eight areas from both a societal and a personal perspective. Hopefully, this format will provide a more viable basis for improving the quality of life.

It seems clear that quality of life needs to be studied in all eight areas to be

Table 3.5
Quality of Life as a Field of Study—Database Format

	Societal		Personal	
Areas	Improvements	Treatments	Improvements	Treatments
Government				
Health				
Work				
Education				
Remote				
Access				
Recreation				
Protection				
Provisions				

Source: Author.

understood fully. The researcher who studies it in only one area is likely to draw biased, thus faulty, conclusions. In addition, clinicians from different areas are often needed to work together as a team in order to improve the quality of a person's life. Specialists working alone quite often cannot achieve the same beneficial results.

Moreover, the quality of life of individuals as a whole is affected by different specifics. The quality of life of a "holy man" may be satisfactory as long as he can practice his religious rituals all day, despite the fact that he is barely able to survive. The dilettante may need to satisfy many different personal desires to be content with the quality of his or her life. And almost anyone can have quality of life destroyed by pain. It seems as though there is a great deal to be gained by viewing quality of life from a more holistic perspective. Most important, people working to improve the quality of life in different ways can have a common focus and understanding that enable them to share and cooperate with each other to become more effective in helping people.

The knowledge base in the field of study consists of the entries that would be in Table 3.5 at any given point in time. The entries in the table should be kept up-to-date to reflect the latest information on improvements in the eight categories and treatments that can achieve the improvements. Whenever a needed improvement is identified, the treatments that can achieve it are retrieved for consideration, including other relevant information such as side effects, costs of treatments, and instructions for administering treatments. The improvement and treatment entries can be expanded as needed. When there are no treatments to achieve an improvement or available treatments are inadequate, the need for further research and development is indicated. Specialists can improve their effectiveness by consulting data in the knowledge base outside their areas of expertise to obtain information relevant to the problem they are attempting to solve at the time. This can only make them more effective problem solvers.

Although some of the listings in guideline Tables 3.3 and 3.4 might serve as appropriate entries in the initial knowledge base, many of them are too general for practical purposes. In addition, the listings are by no means comprehensive enough. An intensive survey was not conducted in each of the eight areas. Moreover, treatments proposed as prescriptions for achieving particular improvements should be rated in some way indicating their proven effectiveness. Theory must be distinguished from fact. And fact must be supported by research evidence. Hopefully, the knowledge base stored in the proposed format will provide a foundation for more effective decision making and research.

DECISION MAKING AND RESEARCH

Two most important uses of the quality of life guidelines and database are decision making and research. Research improves know-how. Decision making puts know-how into practice to benefit people. To facilitate improvement of quality of life, it is necessary to advance the database continually through research and development so that decision making will increasingly become more effective. Decisions founded on the database are evidence-based decisions. Evidence-based decisions are well-informed decisions that are more likely to be accurate and effective than decisions engendered by political pressure, special interest groups, or the subjective judgment of administrators.

Businesses and industries that do not base decisions on objective evidence soon go bankrupt. Many bureaucracies endure even though they do not base decisions on objective evidence and fail to be effective and efficient—for instance, public schools, hospitals, and government agencies. Governments rarely go bankrupt. Rather, most often the coffers are replenished by raising taxes. In a democracy, inefficient, ineffective governments can be voted out of office. Totalitarian governments may not be. Governments in general have no compelling incentive to curb waste. And government officials are frequently seduced and compromised by offers of money or favors. There is always free enterprise in graft regardless of the form of government. Decisions are frequently for sale.

Evidence-Based Decision Making

If quality of life is to be improved, then decisions germane to quality of life need to be based on the research evidence provided in the quality of life database. This ensures not only that the decisions will be evidence based but, in addition, that a wide range of pertinent variables will have been considered. One simply identifies the improvement sought in the database and then considers possible treatments in all of the eight areas of the guidelines before deciding on a treatment plan. Absence of needed data in the database often indicates a need for further research.

Critical decisions are made in all eight quality of life areas delineated in the guidelines, both on a personal level and on a societal level.

On a Personal Level. In the area of government, people decide to improve their knowledge of the law and their legal entitlements. They also decide on the instruction (treatment) they will obtain to learn about their legal entitlements. For example, people enroll in seminars that instruct them on their legal entitlements pertinent to retirement and estate planning. In the area of health, people may decide to improve their strength and endurance. The treatments many decide among may be the gym or health club to join and the exercise regimen to adopt. When people are seriously ailing, the improvement they seek is almost automatically decided: recovery. At such times, they decide how they will treat their ailments. They may decide to try an over-the-counter drug or to consult their pharmacist or family doctor for a treatment prescription. In the area of work, an improvement people may decide to pursue is an increase in income. The treatment they might decide to employ is moonlighting. In the area of education, an improvement people might decide to pursue is competency in using the computer. The treatment they might decide on is enrollment in a computer course. In the area of remote access, an improvement people might decide to pursue is learning to speak Japanese. The treatment they might decide on is taking courses that teach people to speak Japanese. In the area of recreation, an improvement people might decide to pursue is greater competence in playing tennis. The treatment they might decide on is tennis lessons. In the area of protection, an improvement a breadwinner might decide to pursue is increased financial security for his or her children. The treatment he or she might decide on is increasing his or her life insurance. In the area of provisions, an improvement people might decide to pursue is moving to a better neighborhood. The treatment they might decide on is calling a real estate agent.

On the Societal Level. In the area of government, public opinion polls might reveal that people want a reduction in the tenure of elected officials. The treatment that might be decided on is the enactment of a law that imposes term limits. In the area of health, an improvement that might be decided on is improvement in physical fitness. A treatment that might be decided on is a mandatory exercise regimen in all public schools. In the area of work, an improvement that might be decided on is an increase in the rate of employment. A treatment that might be chosen is the attraction of industry to the community. In the area of education, an improvement that might be decided on is an increase in the literacy rate. The treatment that might be decided on is increased reading instruction in public schools. In the area of remote access, an improvement that might be decided on is an increase in the importing of foreign goods. The treatment that might be chosen is a reduction in import taxes. In the area of recreation, an improvement that might be decided on is an increase in the rate of people's involvement in recreational activities. A treatment that might be decided on is an increase in recreational activities in public parks. In the area of protection, an improvement that might be decided on is a reduction in automobile accidents. The treatment that might be decided on is a marked increase in penalties for speeding and reckless driving. In the area of provisions, an improvement that might be decided on is a decrease in slums. The treatment that might be decided on is the replacement of slum tenements with modern dwellings. And of course, as stressed in Chapter 1, although a desired improvement may be in one of the eight areas, it may be beneficial to employ treatments in a number of areas to get the best results. To improve the quality of life of a disabled person, it may be necessary to change his employment (work), train him for his new job (education), administer medication (health), move him to a nurturant residential facility (provisions), and so on.

Research

Too much of the research on quality of life has been fragmentary, disjointed, peripheral, and trivial, as shown in Chapter 2. This is partially because there has been no organizing frame of reference. The quality of life guidelines provide a heuristic framework that can be used to identify (1) significant improvements that are needed to improve quality of life and (2) treatments that can be tested to achieve the improvements. The following are illustrations of how the guidelines might be used to generate germane research.

Societal Research. The following are examples of quality of life improvements inferred from the guidelines for improving quality of life in a society as well as hypotheses that might be tested to effect improvements in society.

Factors	Needed Improvements
Government	*Existing state*: 60% of the eligible citizens vote.
	Desired state: 95% of the eligible citizens vote.
Health	*Existing state*: 20% of the population die from controllable diseases.
	Desired state: 2% of the population dies from controllable diseases.
Work	*Existing state*: 30% of the population's earnings are below the poverty line.
	Desired state: 5% of the population's earnings are below the poverty line.
Education	*Existing state*: 20% of the population are illiterate.
	Desired state: 1% of the population is illiterate.
Remote Access	*Existing state*: 25% of the population can speak foreign languages.
	Desired state: 90% of the population can speak foreign languages.
Recreation	*Existing state*: Public recreation facilities are used 60% of the time they are made available.
	Desired state: Public recreation facilities are used 90% of the time they are made available.

Quality of Life as a Field of Study

Protection   *Existing state*: There are 1,500 crimes reported per year.

                Desired state: There are 300 crimes reported per year.

Provisions   *Existing state*: Imports are $5,925,000 per year.

                Desired state: Imports are $30,000,000 per year.

The following are hypotheses inferred from the guidelines that might be tested to improve the quality of life in societies.

Factors	Treatment Hypotheses
Government	Enacting and enforcing laws that facilitate voter registration will increase the proportion of citizens that vote.
Health	Increasing the number of doctors per capita in economically depressed areas will decrease the number of deaths per capita from controllable diseases.
Work:	Increasing the number of high-tech manufacturing jobs will increase average purchasing power per capita.
Education	Increasing the number of teachers per capita will reduce the percentage of illiterates.
Remote Access	Reducing import tariffs will increase the purchase of foreign commodities.
Recreation	Increasing the number of recreational contests will increase citizen participation in recreational activities.
Protection	Increasing the number of policemen in economically depressed areas will decrease the crime rate.
Provisions	Decreasing the cost per residential unit will reduce the average number of people per room in the population.

Personal Research. The following are examples of quality of life improvements inferred from the guidelines for improving the quality of life of individuals as well as hypotheses that might be tested to improve quality of life on a personal level. The improvements and hypotheses cited may seem unusual because research is not often dealt with on, or couched in a, personal context. However, if we are to improve the quality of life of individuals, it is important that they begin to understand how research and research results can benefit them personally.

Factors	Needed Improvements
Government	*Existing state*: I am unaware of my social security entitlements.
	Desired state: I understand my social security entitlements.
Health	*Existing state*: My depression is interfering with my work and family responsibilities.
	Desired state: My depression no longer interferes with my work and family responsibilities.

Work	*Existing state*: I am dissatisfied with my present job.
	Desired state: I work at a job I find satisfying.
Education	*Existing state*: John can't speak a foreign language.
	Desired state: John can speak Spanish.
Remote Access	*Existing state*: I have not been to the Orient.
	Desired state: I visit China and Japan.
Recreation	*Existing state*: I am bored on the weekends.
	Desiredstate: I adopt a weekend hobby.
Protection	*Existing state*: My family is not provided for after my death.
	Desired state: I own sufficient life insurance to ensure my family's subsistence after my death.
Provisions	*Existing state*: My apartment is inadequate for my family.
	Desired state: I own a home sufficient for my family.

The following treatment hypotheses were inferred from the guidelines for each factor.

Factors	Treatment Hypotheses
Government	Instruction on how to take advantage of legal entitlements will increase the acquisition and access of individuals who received the instruction.
Health	Instruction on the use of over-the-counter headache remedies will decrease the number of people who seek doctors' prescriptions for headache remedies.
Work	Instruction on how to make career advancements will increase the income of those who received the instruction.
Education	Instruction on how to fulfill aspirations within necessary constraints will decrease frustration in individuals who received the instruction.
Remote Access	Instruction on travel opportunities will increase the travel of individuals who received the instruction.
Recreation	Instruction on available recreational opportunities will increase the enjoyment of individuals who received the instruction.
Protection	Instruction on how to protect property will decrease the loss of personal property of individuals who received the instruction.
Provisions	Instruction on how to shop wisely will increase the acquisition of goods and services of individuals who received the instruction.

Of course, experts in each of the eight areas would be able to derive more germane and significant needed improvements and hypotheses.

Having determined the type of research information that needs to be included in the guidelines and database, it is now necessary to consider how to most effectively obtain the data through scientific research.

Chapter 4

Pursuing Improvements Scientifically

The primary question addressed in this chapter is, How can science be employed to improve the quality of life? To answer this question, the rules of science and the scientific method will be reviewed to investigate their relevance to important quality of life issues. In addition, the traditional scientific method will be revisited, and suggestions will be made for adapting it to more effectively solve quality of life problems.

The focus will be on the essence of scientific inquiry more than on technical issues such as methods of sampling, observation, and data analysis. This focus on fundamentals may result in my discussing issues the reader is already familiar with, and some contemporary intellectuals may disagree with my views of science. This can't be helped. Hopefully, I will be interpreting basic issues in a new and constructive way that will make quality of life research more productive, eventually adding more advanced, accurate, and essential data to the quality of life knowledge base.

It seems as though to claim that one is establishing facts scientifically hypotheses must be asserted about future events, the hypotheses must be tested by means of objective observation, and the conclusions must be generalizable. The fact that is ascertained is whether the hypotheses are valid or invalid for particular populations or whether further research is needed to ascertain the validity of the hypotheses. In any event, science requires at the very minimum (1) producing generalizable conclusions (2) by testing hypotheses (3) by means of objective observations. These seem to be the three most fundamental rules or requirements of science.

GENERALIZING

Science recognizes the importance of generalizability to knowledge and human existence. Since no two events are identical, if people were unable to generalize, they would not be able to apply what they know to new situations. Every situation encountered would be totally unfamiliar, and learning would be to no avail. Transfer of knowledge depends on the ability to generalize.

As the reader may know, the most fundamental means of generalizing is through categorizing, which, as indicated, is a matter of identifying entities with common characteristics and designating a word label for them. Doctors diagnose diseases through categorization. The name of a disease is a label for a category. The common characteristics defining the category are the symptoms of the disease. People are diagnosed as having the disease when their symptoms match the symptoms of the disease. Furthermore, knowledge of categories enables people to deal with the novel or unfamiliar. For instance, suppose a person who knows the characteristics of "poisonous snakes" is confronted by a strange snake. He or she would be able to tell whether the snake could poison him or her and act accordingly.

To generalize further, categories are related to one another to describe *how* events are related and to explain *why* they are related. Models, taxonomies, class inclusion and task analysis hierarchies, propositions, and scientific laws describe how categories and events are related. Theories explain why they are related. These schemes for organizing and generalizing knowledge are among the most valuable intellectual tools that humans have fashioned, although they may not be as conspicuous or celebrated as the discoveries and inventions they give rise to.

Generalizing accurately is essential to quality of life research and decision making. Either undergeneralizing or overgeneralizing can have dire consequences for people. The people or population to whom researchers may want to generalize are categorically defined by enumerating their attributes. Since research conclusions are not ubiquitously generalizable, researchers are obliged to define the population they propose to generalize their findings to beforehand as part of their research design. They are further obliged to conduct their research on the entire population or a representative sample of the population so that they can legitimately generalize their findings to their intended population. Otherwise, the usefulness of their findings is seriously compromised. Undergeneralizing their findings results in the findings being applied to too few people; overgeneralizing the findings results in the findings being applied to too many people. In either case, a research error is being made that can critically affect people's lives when findings are applied in decision making.

Clinicians who undergeneralize when diagnosing people to identify the improvement to be pursued will tend to include too few people in a diagnostic category. For instance, when considering the diagnosis of an unfamiliar disease, they will tend not to diagnose people as having the disease who actually have it. Clinicians who overgeneralize will include too many people in a diagnostic category. In diagnosing clinical depression they will diagnose people as suffering from the disorder who are not suffering from the disorder.

Clinicians who undergeneralize when deciding on a treatment will tend to underprescribe the treatment. Consequently, people who can benefit from the treatment will not receive it. This can occur when a clinician has a low opinion of a particular treatment. Clinicians who overgeneralize when deciding on a treatment will tend to overprescribe the treatment. People who cannot benefit from the treatment will tend to be given the treatment. This can happen when a clinician is oversold on a treatment. These errors are more than logical possibilities; they are consequences of bias, a characteristic of all living human beings. Research has shown that a doctor's interpretation of a patient's recovery is correlated with the confidence he or she has in the effectiveness of the medication that has been prescribed. So clinicians do develop biases toward the medications they prescribe; and this can in turn affect their interpretation of the effectiveness of the medications, the frequency with which they prescribe the medications, and the number of people who receive the medications.

In research, overgeneralization tends to be more pervasive than undergeneralization. The hypotheses researchers choose to test are often hypotheses they believe are valid and of widespread significance. They, like clinicians, tend to interpret findings in favor of their biases. They often tend to argue for the validity of their hypotheses in their conclusions even when the findings do not support it. Moreover, they tend to overgeneralize their conclusions beyond the population studied.

Overgeneralization seems to be a form of aggrandizement. It seems that people equate importance to a greater degree of generalizability. For example, a medication that benefits a greater number of people is usually considered more important. Similarly, a disease that affects a greater number of people is often considered more important. The disease is more likely to receive pervasive attention and more research funding.

Generalizability of conclusions is especially important when research is being done to determine whether a particular improvement should be pursued as a matter of public policy and whether a particular treatment should be approved for public adoption. Research findings that are not publicly generalizable are of little value in setting public policy. Still, to be of value, research findings need not be generalizable to all humans. Research findings seldom, if ever, apply to everyone unconditionally.

To grasp the problem, it is helpful to understand that all three of the following statements are in general true.

1. A person is like all other people; all people have four-chambered hearts.
2. A person is like some other people; some speak English, some Spanish.
3. A person is like no other person; no two sets of fingerprints are identical.

It appears that the greatest generalizability can be obtained with respect to physical phenomena: Every living person has a heart, and almost all human hearts are four chambered. However, I imagine there may be rare exceptions. Social findings appear to have more limited generalizability. Social groups are different from one another, yet there is considerable generalizability within social groups with respect to language, beliefs, and rituals. Most Americans speak English, endorse the Bill of Rights, and stand when singing "The Star-Spangled Banner." Research findings generalizable to Americans might have significant value to Americans, despite the fact that they may not apply to anyone else.

People who emphasize the fact that a person is like no other person emphasize individual differences and idiosyncrasies. In the extreme, they hold that people are so different from one another that what applies to one person cannot possibly apply to another. To help people, one must look at them as individuals, as unique entities. Although it is important to consider people's idiosyncrasies, if we are to help people, we must also be able to apply validated generalities to people, whether the generalities apply to all people or only to the social group to which the person belongs. Many factors that affect quality of life are social factors, for example, education and government.

Since there are no panaceas, successful treatments cannot be expected to have unlimited generalizability. The ultimate aim is to have a number of treatments for a type of ailment such as headaches, so that the public may have access to all of them, with the advice and consent of experts. Only then can we expect to be able to bring relief to a maximum number of people, given individual differences in treatment effectiveness and tolerances across people.

One of the most frequent and problematic forms of overgeneralization emanates from laboratory research. Research on laboratory animals is all too generously generalized to humans, with very serious consequences. Such overgeneralizations result in approving medications for public use that are harmful—for example, thalidomide, which is now being reevaluated. At other times, overgeneralizations result in people becoming unnecessarily alarmed about a product, for example, the saccharin scare.

Overgeneralizing from infrahumans to humans can be reduced if funding agencies that provide money for research on laboratory animals consider, before making such grants, the implications for making decisions about humans. If the results of a study cannot be directly and safely generalized and applied to humans, the study is of little value and can be dangerous. It would be a far better practice for funding agencies to identify important decisions that need to be made to improve the quality of human life and fund research that will enable those decisions to be made.

Another frequent cause of overgeneralization is insufficient sample size. Probably the most insidious and misleading form is the testimonial of an actor in a commercial to the effect that a product he or she tried is curative. Although the testimonial may not be directly asserted as a research finding, it suggests that the public use the product based on the testimony of one individual. Products are also tried based on a sample of one when a person tries a product owing to a recommendation of a friend. The problem of basing a conclusion on one case is sometimes highlighted with the statement: All Indians walk in a single file—at least the one I saw did.

Even when generalizations are based on a larger number of cases, the number of cases is quite often too few to generalize to the public. It takes a very large sample size to enable a generalization to be made to the public. And it takes vast sums of money to obtain such a large sample. Still, those who want to sell an idea or a product to the public will make their claims based on an insufficient sample size. This practice has become more problematical as more and more politicians and manufacturers hire pollsters to find evidence to support the claims they want to convey to the public. In most cases, there is only one permissible finding, the finding that supports the claim of the people who hire the pollster.

Clinical findings are also frequently overgeneralized. Clinicians interested in research analyze the data they accumulate in their clients' files and tend to be too liberal in generalizing their findings from the few clients they have seen in their practices. Although clinical findings are valuable, they are seldom generalizable to the public. Clinical findings are most valuable when used as a basis for deriving important hypotheses to be tested on samples of adequate size. It is also important to realize that clinical testing instruments designed to diagnose personal problems and to assess progress toward recovery may not be adequate for quality of life research that is intended to be generalizable to the public.

OBJECTIVITY

Objectivity is a rule of science that is given a number of definitions. A common definition in science textbooks is the absence of subjective judgment. I prefer a people-oriented operational definition that is germane to quality of life research.

Objectivity is crucial to people because it is the basis for sharing. Without objective agreement, people would not be able to share ideas, profit from one another's experiences, or cooperate. Civilization as we now know it would be impossible. To avail ourselves of these benefits, there must be agreement among people about common experiences. Dictionaries manifest our objective agreement on linguistic symbols and what they denote. By using the words in the dictionary, people can share perceptions, which is prerequisite to other kinds of sharing.

The following definition of objectivity is proffered because it highlights the importance of objectivity as a prerequisite to sharing: Objectivity is achieved when a group of qualified observers agree on an observation. To be qualified the observers learn and follow the same rules for making observations. For example, to observe the height of a horse objectively, a group of people learn the rules of using a metric ruler to observe height. They then follow the rules to measure the horse's height. To the extent that they can agree on the horse's height their observation is objective. If 100 percent of them agree on the horse's height, there is greater objectivity than if only 80 percent of them can agree.

Once objective agreement has been obtained about an event, it can be conveyed as public knowledge with greater probability that it is accurate than the subjective impression of any individual. This, I suggest, is the primary significance of objectivity to the human race.

Moreover, because objective evidence is likely to be more accurate, decisions based on objective evidence are likely to be more accurate. On the societal level, any public policy decision to authorize a product or practice for public use is more apt to be accurate as warranted. And any personal decision based on objective evidence is more likely to be accurate than a decision based on one's own personal opinion or the personal opinion of an observer. The fact that individuals may be under no obligation to base their personal decisions on objective evidence is beside the point. One of the main problems today is that public policy is too often determined by vested interests and dominant political figures or factions rather than by objective evidence. With respect to the proposed guidelines for improving the quality of life, objective evidence needs to be used to identify needed improvements and to assess treatment effects and side effects as well as costs of administering treatments.

Now, in analyzing the quality of life instruments reviewed in Chapter 2 and comments about them, a great deal of variation and confusion about objectivity was evident. Instruments were reviewed in two major categories, observation instruments and self-report instruments. It appears that some instrument specialists regard observations as objective and self-reports as subjective. From my perspective, objective information can be obtained using self-reports if there is agreement among those giving the self-reports on a particular issue. And objective information can be obtained from observations only if there is agreement among observers on a particular issue.

Objectivity depends on agreement among people applying the same rules to make a decision whether the decision is based on observations or self-reports. Objectivity of personal opinions is estimated quite often in marketing when public opinion polls are used to determine whether a practice or product is appealing. For example, a headache remedy manufacturer might survey a representative sample of the public who have used its product according to the rules published on the container to get relief from headaches. Objectivity among observers would be obtained to determine whether people observing a phenomenon using the same rules agreed in their observations of the phenomenon. For instance, a survey of dermatologists prescribing a cream to cure a fungal infection might be made to determine whether the dermatologists agreed that the cream cured their patients' fungal infections.

It is the practice to regard scores given on multiple-choice and true-false tests as objective because a scoring key is used to score them and anyone using the key to obtain a score will derive the same score. However, shortcuts do not always produce objective results. For instance, using a scoring key to assess the learning of math can be expected to produce objective results because in math there most often is only one correct answer, and people have little difficulty in agreeing on the correct answer. On the other hand, using a scoring key to grade English compositions could not be expected to produce objective results. There are great variations of opinion among English composition experts on what constitutes good English composition. It is not uncommon for one English teacher to assign a high grade to an English composition, while another teacher assigns a low grade to the composition. Instruments that provide a scoring key cannot be presumed to produce objective results. In general, instrumentation is always an issue when objectivity is being considered.

Hard-nosed scientists primarily from the physical sciences have traditionally claimed that for observations to be objective they need to be made by means of measurement. This, I suggest, is unnecessarily rigid. If qualified observers following the same rules of observation agree on an observation, objectivity has been established whether the observation has been made by means of quantitative or qualitative observation. To make objective observations quantitatively, rules of measurement are used. To make qualitative observations, rules of common language are used.

Many important scientific discoveries have been made by means of qualitative observation, for example, William Harvey's discoveries about the circulation of the blood. He observed blood coursing through blood vessels in rhythm to the heartbeat and communicated his observations and findings to others through common language. Objectivity was established when others confirmed his observations.

The ability to measure characteristics of a phenomenon such as blood flow develops with research over time. And the availability of numerous measuring instruments to measure a phenomenon is a sign of a more advanced discipline. Since Harvey first discovered the circulatory system, more and more ways of measuring characteristics of the blood have evolved with research and development over many years. Claiming that researchers like Harvey who make discoveries without using measurement are not scientists is unnecessarily restrictive and counterproductive.

Although objectivity can be achieved by means of qualitative as well as quantitative observation, those who do qualitative research, such as ethnographers, do not as often report or achieve objective agreement. This is partly because objectivity may be more readily established through measurement. Still, in many studies I have read that record qualitative information, no testing for objectivity has occurred. Such research has no claim to being scientific research. For qualitative research to be more valuable to the public, objectivity must be achieved.

It is also artificial and counterproductive to continue the schism between qualitative and quantitative research that arises from abstract arguments among qualitative and quantitative researchers. Quality and quantity are not independent of one another. For instance, the quality "water" is defined in terms of the quantities of three variables, hydrogen, oxygen, and temperature: two parts of hydrogen to one part of oxygen, between 32 and 212 degrees Fahrenheit. Raising the temperature to 212 degrees changes the quality to steam. Lowering the temperature below 32 degrees changes the quality to ice. Quantitative variables vary in degree as temperature does. Qualitative variables vary in kind, as water is categorically different from steam because of their different composition of variables. In the long run, productive research cannot be conducted conceiving of quantity and quality as separate issues.

It is also important to realize that in observing change in quality of life both quantitative and qualitative observations can, are, and should be made as appropriate. A doctor determines whether a patient is recovering from an illness quantitatively by measuring the patient's body temperature, when it has been elevated. A doctor also considers qualitative changes. For example, he or she determines whether a patient who had symptoms of a disease, say, "pneumonia," no longer has the symptoms of the disease and can be pronounced "well." In diagnosis, observing qualitative transitions from one state to another is as necessary as observing quantitative changes in symptoms.

All of the controversy over the definition of quality of life notwithstanding, it is clear that quality of life is a qualitative or categorical variable composed of and defined in terms of a number of variables. Quality of life is not a quantitative variable. The issue in question is how to define quality of life in terms of other variables. Definitions were offered in Chapter 3 for consideration.

Objectivity depends on reliability. Repeated observations that are made of the same phenomenon using an unreliable instrument will be highly variable. Consequently, it will be unlikely that observations of the same phenomenon by different observers using the instrument will be in agreement. Making observations with an unreliable instrument is like measuring the width of a table with a rubber yardstick that keeps expanding and contracting. So since science requires objective observation, instrument reliability is important because it enables objective observation.

Although objectivity depends on reliability, validity is independent of both reliability and objectivity. An observation instrument can yield both reliable and objective observations and be useless because it is not valid; that is, it does not observe the variable it was constructed to observe. What good are objective and reliable observations of the wrong phenomenon? To establish the validity of an instrument scientifically, it would, strictly speaking, be necessary to establish predictive validity by means of hypothesis testing. For example, if the validity of a functional ability instrument were to be established scientifically, it would be necessary to test the hypothesis that people who score high on the instrument will be able to perform tasks that require greater functional ability than those scoring low on the instrument. But hypothesis testing most often requires the expenditure of a sizable amount of time and other resources, because it is necessary to wait until a future time to determine whether or not a hypothesis is valid. As a result, it has become customary to establish instrument validity more expeditiously by establishing face, logical, construct, content, and concurrent validity, as may be considered appropriate. Still, predictive validation is the means of validating instruments scientifically.

It is interesting to note in passing that because validity and reliability are inherent properties of instruments, they are dealt with intensively in textbook and professional discourses on instrumentation. In contrast, because objectivity is only sometimes an inherent property of instruments, it is often given short shrift. Ultimately, we are concerned with the validity, reliability, and objectivity of observations—all three, however the observations may have been obtained.

To understand objective judgment fully, it is necessary to understand subjective judgment as well. Many of the decisions people make are based on personal preferences and do not require objective evidence. The clothes people buy and choose for their wardrobes, the mates and friends people choose, the names people choose for their children, the rites people practice, the goals people pursue, and many other choices people make are based on personal preferences. One of the advantages of living in a free society is that people are free to do as they please within minimal legal limits. People are free to marry whom they choose, eat what they choose, read what they choose, worship as they choose, and so on. And objective evidence is not required or often sought to make the choices.

A grave mistake teachers make in modern societies, especially science teachers, is to lead students to believe that facts can only be established by means of objective scientific evidence. This deprives youth of the rich, lavish experiences that can be enjoyed in life. There is subjective experience just as sure as there is objective experience. And much of the enjoyment people get out of life is from subjective experience. Objectivity enables people to share scientific advancements. Subjectivity enables people to experience personal satisfaction, sensual pleasure, and self-fulfillment. Life can become sterile and vapid when objective experience is overemphasized. Being able to exercise personal preference enriches quality of life whether it is allowed by law or by a benevolent dictator.

One way people are allowed to express their personal preferences in free societies is in deciding whether to accept or reject proposed treatments. For example, the Massachusetts Supreme Court in 1982 reaffirmed the doctrine of informed consent (Harish 1982). The following highlight important provisions:

- The patient, not the physician, determines the direction in which the patients's best interests lie.
- Every competent adult has the right to forego treatment, even cure, if the patient believes the treatment holds intolerable consequences or risks. The patient may do so no matter how foolhardy the medical profession may consider that decision.
- For the patient to make the choice, the patient must be advised as to the available options for treatments and risks attendant to each option.

Thus, the medical profession must provide objective evidence of treatment options and their risks as a basis for patients to make personal decisions.

People must be allowed to make personal decisions for a number of reasons. Evidence may be ambiguous. The evidence obtained may have demonstrated some degree of improvement, but the improvement may be subjectively judged too little and/or the side effects too damaging. Furthermore, issues other than treatment effectiveness can determine a decision. Treatments may not be used because they are too costly, because it may be illegal to impose them on others without their permission, or because they are considered immoral.

Still, if people are to be able to make enlightened decisions to improve the quality of their lives, they need to learn to identify and consider objective evidence. If they are on a diet to lose weight, they should be able to keep track of the calories and fat they consume by reading the labels of food products. They should also be aware of acceptable cholesterol levels in their blood and know how the fat content listed on food labels may affect those levels. To save money, they should know how to interpret the cost per ounce of food products displayed on supermarket shelves and how to construct and interpret a household budget. They should also understand how to use and read a thermometer to determine body temperature and to interpret their findings. Most important, they should know how to seek the advice of a professional when they have difficulty interpreting objective evidence and to get a second and perhaps a third professional opinion when they are in doubt.

Although clinicians are obliged to take the aspirations and desires of their clientele into account, they should guard against being misled by their subjective judgments. It is not uncommon for a client to attempt to persuade a clinician to prescribe a particular treatment. Giving clients what they prefer does not absolve clinicians from being responsible for the effects of the treatments they prescribe and administer or from making mistaken diagnoses. Clinicians must base their decisions on objective evidence. Otherwise, their decisions are less likely to be effective. In addition, they are not defensible scientifically or in a court of law. Clinicians must stand ready to defend that their diagnoses are based on objective evidence and that there is objective evidence supporting the treatments they prescribe. They cannot afford to do otherwise, regardless of the subjective judgments of their clients. Furthermore, they should not be dissuaded by the arguments of testing specialists who believe that quality of life is entirely a personal matter expressed through personal opinion. Even when it is defensible for clients to choose among treatment options, as indicated, clinicians are obliged, if not legally required, to provide objective evidence of diagnoses as well as treatment options and their risks.

TESTING HYPOTHESES

The scientific rule that facts are established by means of hypothesis testing suggests that facts are not established retrospectively. Ex post facto explanations are not allowed. Furthermore, a major value of history is that pregnant, meaningful, cogent hypotheses can be derived from historical accounts to be tested in the future. Those who attend horse races buy racing forms to check the past performances of horses to determine which horses to bet on in coming races. As a basis for selecting treatments they hypothesize will remedy their patients' ailments in the future, doctors query patients about their past experiences to find the causes of their present ailments. In the scientific context, history is the best resource for deriving enlightened hypotheses that are likely to be validated when tested in the future.

In essence, scientific facts are established when hypotheses are tested by means of objective observation and the findings are generalizable over time and space. As mentioned earlier, the means of testing hypotheses scientifically has come to be known as the scientific method, which varies considerably depending on who is describing it. Although most people have some knowledge of the scientific method, it needs to be reviewed to clarify its functions and limitations. The scientific method can be boiled down to four basic stages.

Stage 1: Deriving hypotheses indicating the information sought.

Stage 2: Deriving procedures to test the hypotheses including procedures for describing and observing the population being investigated, objectively observing and describing the variables of concern, and determining the relationships between the variables if relationships are being determined.

Stage 3: Following the derived procedures to test the hypotheses.

Stage 4: Interpreting the findings to draw conclusions and to make recommendations for theory and/or practice as well as for future studies.

The scientific method in its purest form is a knowledge-producing rather than a problem-solving or improvement-achieving method. It produces knowledge on four levels, commensurate with the four purposes of science often cited: the control, explanation, prediction, and description of events.

Levels of Knowledge

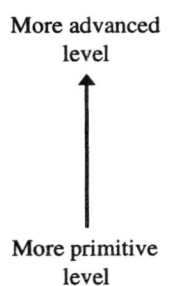

More advanced level
↑
|
|
More primitive level

1. Control level: generates knowledge about the application of treatments to control events

2. Explanatory level: generates explanations of cause-effect relationships (which enhances theory development)

3. Predictive level: generates knowledge of how to predict one event from another (or others)

4. Descriptive level: generates knowledge of how to describe events

The four levels form a hierarchy in which knowledge at a more advanced level implies knowledge at all lower levels. First, this means that if individuals are able to control, they can also explain cause-effect, predict, and describe. For instance, if they are able to control desired learning through instruction, they know that the instructional method causes the learning, they can predict that the desired learning will occur when they use the instructional method, and they can describe the desired learning, the instructional method, and their relationships. Second, if individuals can explain cause-effect, they can predict and describe. For example, if they know that hurricanes cause damage, they can predict that when a hurricane comes, damage will occur; and they can describe hurricanes and the damage they effect. However, knowledge of cause-effect does not imply control of the causal agent. We cannot control hurricanes as yet. Still, some control is implied. We can control our own actions to flee from a hurricane when we predict one is coming. Third, if individuals can predict, they can also describe. For example, the ability to predict that night will follow day implies the ability to describe night, day, and their relationship. However, the ability to predict does not imply knowledge of cause-effect or the ability to control. Knowing that night follows day does not imply that day causes night. Nor does it imply that day can be manipulated to control night. And fourth, the ability to describe does not imply the ability to predict, explain, or control. However, being able to describe events provides a basis for conducting research that can lead to prediction, explanation, and control.

The format presented at the end of Chapter 3 that identifies quality of life as a field of study focuses on and abets the elevation of quality of life knowledge to the control level by emphasizing the importance of finding treatments to control improvements. Ultimately, the mission is to improve quality of life by controlling through treatments those factors that affect quality of life. The mission is to do more than explain why certain factors affect quality of life, to do more than predict improvement in quality of life, and to do more than describe indicators of quality of life. The mission is to control the quality of people's lives so that people can at will manipulate the factors necessary to elevate their lives. To date, a great deal of quality of life research has been on the descriptive and predictive levels—for example, the construction of instruments to describe functional ability and the search for indicators that can predict quality of life.

Now, the scientific method is effective if the researcher's objective is to apply it to increase the fund of knowledge in some way. It is less effective if the purpose is to work to progressively converge on the achievement of an improvement. The scientific method is essentially a divergent method in that the testing of one hypothesis almost inevitably reveals the need to test a number of additional hypotheses in future studies. It is in this way that repeated applications of the scientific method expands knowledge. Researchers who conscientiously apply the scientific method are obliged to list hypotheses recommended for future studies. Each study not only adds to the existing fund of knowledge; it suggests hypotheses that need to be tested to expand the knowledge base further.

Pursuing Improvements Scientifically 103

Figure 4.1
The Control Cycle

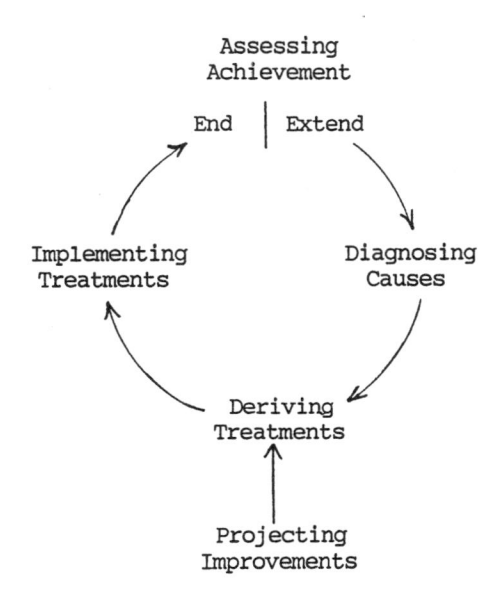

```
                    Assessing
                    Achievement
              End  |  Extend
               ↗           ↘
    Implementing           Diagnosing
    Treatments             Causes
          ↖           ↙
            Deriving
            Treatments
               ↑
            Projecting
            Improvements
```

Source: Author.

CONTROLLING THE ACHIEVEMENT OF IMPROVEMENTS

The scientific method is not an effective procedure for progressively converging on the achievement of an improvement. Rather, converging on an improvement requires an application of the feedback loop, which has come to be regarded as the basic unit of analysis for adaptive human behavior. Following this mode, once an improvement is identified, behavior is identified that is hypothesized to achieve the improvement, and the behavior is then tested to see if it does. If unsuccessful clues are gleaned based on feedback, modified behaviors are hypothesized to achieve the improvement and tested on a next attempt. The procedure continues, attempt after attempt, as the achievement of the improvement is converged on by successive approximation.

A general adaptation of the feedback loop rationale has been developed to facilitate the achievement of improvements. It is called the control cycle (Figure 4.1) because it is a method for controlling the achievement of improvements.

Projecting Improvements

As indicated, improvement is progress from an existing state to a desired state. The existing state represents the starting point; the desired state is the goal to be achieved. Defining improvements in this way enhances both the derivation of treatments and the assessment of achievement. The derivation of treatments is facilitated when it is evident that a treatment must provide movement from a particular starting point to a particular goal. Specifying the goal alone is not sufficient. Assessment of achievement is aided when progress can be followed from a present starting point to a goal.

Deriving Treatments

Deriving treatments involves (1) determining the factors to be controlled to achieve the improvement by analyzing the discrepancy between the existing and desired states; (2) determining constraints that must be met in pursuing the improvement; (3) determining means of controlling the factors that must be controlled; and (4) deriving a treatment for controlling the factors to achieve the improvement. These four steps will be exemplified in Chapter 6. Treatments are not always manipulative interventions. Sometimes, the derived treatment may be to do nothing. Such is the case when a harmless disease is running its course, such as the common cold.

Implementing Treatments

To determine whether or not the treatment achieves the improvement, the treatment must be administered according to specifications. To ensure accurate administration of the treatment, administrations must be monitored to detect and correct variations from specifications before they become too pronounced When treatments are long and complex, rehearsals and other preparations may need to be made before the treatment is administered.

Assessing Achievement

After the treatment is completed, actual outcomes are compared with the desired state to determine the extent to which the desired state or goal was achieved and to assess side effects. Observation of outcomes may require the use of measurement instruments. Accurately comparing actual outcomes to goals may require the use of evaluation and statistical techniques. Also, a system may need to be devised to ensure accurate interpretation of comparisons. For instance, scorers of diving competitions are trained to make interpretations of diving performances against established criteria, and a system is used to approximate objectivity.

If improvement is satisfactory, the process ends; if not, the process extends.

Diagnosing Causes

When improvement is unsatisfactory, an attempt is made to diagnose causes of the failure as a basis for devising a new treatment. First, degree of improvement is considered. If some progress was made and there is evidence that the treatment is controlling the factors that need to be controlled, then an increase in the intensity and/or time of treatment is indicated. Also, slight variations in the means of controlling the factors to be controlled might be tried. If significant progress was not made and there is evidence that the factors that need to be controlled are not being controlled, then new means of controlling the factors need to be tried. If significant progress was not made and the factors to be controlled are being controlled, then there is a need to look for new factors to be controlled, as well as the means of controlling these factors. In the search for a cure for acquired immunodeficiency syndrome (AIDS), for example, some scientists are suggesting that factors other than the human immunodeficiency (HIV) virus are factors that need to be controlled. Through logic, new treatments can be derived from previous achievement that can be predicted to be more effective.

It should become clear that improvements are achieved by identifying factors that need to be controlled to achieve the improvement and then finding the means of controlling the factors to bring about the improvement. When a new means of controlling factors has been derived, an innovative treatment has been achieved. Innovation will be discussed in detail in Chapter 6.

Movement through the cycle continues, cycle after cycle, until by successive approximation satisfactory improvement has been achieved. Of course, the quest may be aborted at any time because priorities change or resources are no longer available.

It would seem that the control cycle would in most cases be a more effective means of improving the quality of life than the scientific method, since the scientific method is designed more to increase knowledge and the control cycle is designed more to control the achievement of improvements.

As one contemplates applying the control cycle to achieve an improvement, it is important to consider key issues that may need to be taken care of by progressing through the following decision-making steps. The issues need to be addressed in a particular order and have implications for types of research and/or development that may need to be conducted in order to apply the control cycle.

Step 1: Establishing the Desired State

Determine whether the desired state has been authoritatively and objectively established by qualified people. If so, move to Step 2. If not, it is necessary to establish the desired state before proceeding. Duly elected legislators are authorized to establish desired states by enacting laws. Objectivity is achieved through agreement among committee members or by majority vote. However, in many instances, desired states are established without authority or objectivity by people whose qualifications are dubious. The type of research done to establish desired states is often referred to as *policy research*. In policy research to establish desired states, the authority and qualifications of those involved in the decision-making process must be established as well as the means of arriving at decisions objectively. The Delphi Technique can be used to establish consensus among authorities without face-to-face contact.

Of course, individuals usually have the authority to establish desired states for themselves within the law. And they usually have the prerogative to rely on their personal opinions or to use objective evidence to come to a decision as they choose. On the other hand, in matters that affect public policy, desired states should be established authoritatively and objectively by qualified people.

Step 2: Observing the Existing State

Determine whether there is an accurate procedure available for observing the existing state. An accurate observation procedure is needed to determine whether improvement is needed in the first place and to assess achievement of improvement.

If there is an accurate procedure, move to Step 3. If not, one needs to be developed. This type of development is often referred to as *instrument development*. Before an observation procedure is ready for use, its accuracy must be demonstrated. This is done by showing that when the instrument is used, it yields valid, reliable, and objective observations. It is beneficial if instruments used to detect improvement are designed to diagnose causes of lack of improvement so that failure can be explained and remedial treatments can be developed more readily. Too many observation instruments indicate achievement without being able to diagnose causes of lack of achievement. For example, people who are not given promotions are often told that their achievement did not warrant promotion, but their specific failures are frequently not diagnosed. Consequently, it is impossible to determine what they need to do to be promoted in the future.

Step 3: Establishing the Need for Improvement

Determine whether improvement is needed in the population of concern by comparing the existing state to the desired state to determine whether there is a discrepancy between the two. If improvement is needed, move to Step 4. If not, there is no need for the contemplated project. There is never any shortage of dissenters, protesters, and advocates for causes. Quite often, however, they are unable to back up their advocacy for needed improvements with objective evidence. Still the admonition "If it ain't broke, don't fix it" is frequently ignored, and money is wasted on unnecessary projects, which sometimes do more harm than good. In the United States in the late 1900s, some people were declaring a health care crisis, while others claimed that the United States had the best health care system in the world. Still others were claiming that it was not so much a health care crisis as a health cost crisis. A needed improvement should be defined and objectively shown to be a discrepancy between an existing and a desired state. In this way the need for the improvement is clear, and there is a sound basis for developing treatments to achieve the improvement as well as for monitoring progress in achieving improvements. Determining discrepancies between desired and existing states is generally referred to as *evaluative research*, whether it is done to determine the need for improvement initially or to assess achievement as a result of treatment.

Step 4: Establishing Treatment Effectiveness

Determine whether there are known treatments that can be hypothesized to induce improvement. If not, move to Step 5. If there are, conduct research to test their effectiveness. This type of research is generally referred to as *experimental research* aimed at controlling effects or achievement of improvements. In experimental research, experimental groups are given treatments hypothesized to induce improvement. Control groups are not given the treatments. Instead, they are given no treatment, a placebo, or an inadequate traditional treatment. In all other ways, the experimental and control groups are made as equivalent as possible so that differences in outcomes can be attributed to treatment differences. In order to test the effectiveness of a treatment, it must be administered according to specifications.

Step 5: Developing Treatments

Determine whether the factors that need to be controlled to induce the improvement can be identified as a basis for deriving a treatment. If not, move to Step 6. If they can be identified, derive a treatment to manipulate the factors so as to induce improvement. This type of project is often referred to as *treatment development*. Explicit instructions for administering treatments must be specified. Otherwise, the treatment cannot be administered and tested accurately. Once a treatment is developed, its effectiveness can be tested by conducting experimental research, as described in Step 4.

Step 6: Identifying Factors to Be Controlled

Determine whether factors that need to be controlled to induce the improvement can be derived. To identify factors that need to be controlled, specific causes of the improvement need to be identified. If causes can be identified, then treatments can be developed to manipulate the causal agents to bring about the improvement. It is the purpose of theories to identify cause-effect relationships. Potential causes of the improvement can often be derived from a review of relevant theories. If possible causal agents can be identified, their causal properties can be tested by conducting *causal-comparative research*. Causal-comparative research is ex post facto research designed to establish causality when experimental research is premature or not feasible. To test the hypothesis that a causal agent will bring about an effect or improvement, groups with the causal agent present are compared to groups with the causal agent absent. If the effect occurs repeatedly in the presence of the causal agent but not in its absence, then it is concluded that the agent is a cause of the effect. If causal agents can be identified, treatments can be selected or developed to manipulate the causal agents in order to bring about the improvement, as explained in Step 5. If causal agents cannot be identified, move to Step 7.

Step 7: Identifying Variables That Predict the Improvement

Determine whether variables that predict the improvement can be identified. When one does not know of causal agents that are theorized to bring about the improvement, *predictive research* is in order. In predictive research, the researcher is determining whether selected predictor variables, which prior research suggests are associated with the improvement, individually or in combination, actually predict the improvement. If predictor variables do predict the improvement, then further causal-comparative research can be conducted to determine whether the predictor variables are causal agents of the improvement. If prior research does not reveal variables associated with the improvement, move to Step 8.

Step 8: Identifying Variables Associated with the Improvement

Determine through *descriptive research* variables that are associated with the improvement. Descriptive research takes many forms. Historical research describes past events. Ethnographic research describes events that have been observed as unobtrusively as possible in their natural settings. Survey research describes the opinions of people often in response to questionnaires. If variables associated with the improvement can be identified, they can be used as predictor variables in predictive studies to determine whether they predict the improvement. It should be pointed out that research conducted to establish desired and existing states is descriptive research, as is evaluative research conducted to describe discrepancies between existing and desired states. Also, instrument development is for the purpose of describing variables, and treatment development ends up with a description of treatment specifications.

The types of research discussed above can be associated with the levels of knowledge presented earlier. Experimental research is associated with the control level because treatments are administered to manipulate causal agents to control outcomes or improvements. Causal-comparative research is associated with the explanatory level because an attempt is made to identify and explain causes of effects. Predictive research is of course associated with the predictive level because an attempt is made to predict outcomes or improvements. And descriptive research is associated with the descriptive level because an attempt is made to describe variables and associations between them.

It is most important in conducting research and development that the purposes and limitations of the type of research be kept in mind. It is not uncommon to find in descriptive studies, where only descriptive conclusions are warranted by the descriptive data, extravagant conclusions explaining cause-effect, warranting the use of treatments to control outcomes or claiming that the data support particular predictions.

Evaluation or evaluative research needs to be used more often in testing treatment effectiveness, especially when a treatment is being considered for public adoption. Too often, treatments are compared to one another, and the treatment that gets the best results is adopted for use. However, it might be that the treatment that produced the best results is not sufficiently effective to be adopted for use or that both treatments are sufficiently effective to be adopted for use. In the latter case, other criteria such as treatment cost might be taken into account before selecting a treatment. Criteria of treatment effectiveness (the desired state) would need to be established and treatment effectiveness would need to be evaluated by comparing the effects of the various treatments to the criteria to make a more cogent decision. Just testing the relative effectiveness of different treatments is usually not sufficient or defensible for adopting treatments for public use. Policy makers need to establish criteria of desirability or minimum criteria of acceptability before testing treatment effectiveness.

It has been mistakenly claimed, based on the traditional views of the physical sciences, that causality can only be established through experimental research. As valuable and precise as experimental research may be, causal-comparative research is now used commonly in social research to establish causality. Experiments often infringe on the rights of human subjects, and the artificially controlled conditions imposed in experiments frequently preclude applying or generalizing the results of an experiment to real-world situations. Consider how the conclusion that cigarette smoking causes lung cancer was arrived at by conducting causal-comparative research.

In general, the researchers contrasted groups of cigarette smokers with groups of nonsmokers. The great majority of the time, they found a higher incidence of lung cancer in the group of smokers than in the group of nonsmokers. The primary control they used was excluding from the sample people who were in poor health. The contrast was replicated many times with much the same results. With each replication, it became more probable that cigarette smoking is a cause of lung cancer. When the level of probability became compelling, the claim was made that cigarette smoking is a cause of lung cancer. And there is hardly an intelligent, informed person alive that is not convinced. According to the laws of probability, there is very little chance that cigarette smoking is not a cause of lung cancer.

Now consider how the research would need to be conducted ideally to satisfy the stringent requirements of experimental research. There would first be an attempt to match the subjects in the experimental group of smokers and the control group of nonsmokers. This might be done by acquiring identical twins at birth and assigning them at random to the experimental and control groups. This is analogous to assigning mice from the same litter to experimental and control groups in medical research. Both the experimental and control groups of children would be held captive in a pollution-free environment so that air pollutants could not be a cause of lung cancer, should it occur. And both groups would eat the same healthy diets and exercise the same amount. Continuing to make all other factors as equal as possible for the experimental and control groups, up to the age of, say, 15, the youths in the experimental group would be forced to smoke, say, two packs of cigarettes a day, while the control group was not allowed to smoke at all. At about age 35, after the experimental group had been smoking two packs a day for 20 years, the two groups would be compared for incidence of lung cancer. If there were a significantly greater incidence of lung cancer in the experimental groups than in the control group, all other things being equal, it would be concluded that cigarette smoking is *the* cause of lung cancer.

As beneficial as the experiment may be in the physical sciences where physical matter can be manipulated as the experimenter chooses, and with laboratory animals where the primary source of complaints comes from the Society for the Prevention of Cruelty to Animals, experimental research will probably continue to be the primary source of establishing causality. On the other hand, researchers working with humans will quite often need to rely on causal-comparative research to establish causality. And those who want to stubbornly maintain that only experiments establish causality may choose to demean the importance of causal-comparative research in establishing causality and continue to believe that cigarette smoking has not been proven to be a cause of lung cancer.

Still, as logical as the experimental format may be, it is in some ways superficial and unrealistic. In the real world, experimental and control groups cannot be made equal as experiments require, and the more complex the phenomenon being studied, the more difficult it is to achieve equality. Since humans are the most complex of creatures, equality is most difficult to achieve. Second, experiments propose that the treatment being used in an experiment can be established as the cause of the outcome. In scientific research, outcomes are caused by multiple factors, not single factors. Cigarette smoking is not *the* cause of lung cancer, and the cigarette manufacturers who claim that cigarettes have not been proven to be the cause of lung cancer are technically correct. It is the interaction of cigarette smoke ingredients with various chemical components of body tissue that causes lung cancer. Proposing single causal agents of effects is naive, even though it may be some time before the causes of lung cancer are completely isolated. To more fully appreciate the notion of multiple causality, consider that the causes of flame are at least oxygen, fuel, and an igniting agent. In science, the law of parsimony requires that necessary and sufficient causes be established for an effect.

The real advantage of the experiment is that it enables manipulative controls to be applied to make the subjects and environmental conditions in experimental and control groups as similar as possible. The disadvantage is that in manipulating subjects researchers may cause them to behave unnaturally, and in manipulating the environment, they might make it considerably different than the subjects' natural environments. In addition, the manipulations may violate the legal rights of the subjects. Manipulative controls cannot be used in causal-comparative or predictive research. Consequently, the concerns of manipulative controls are not present.

The most useful controls that can be used in predictive and causal-comparative research are statistical and population controls. A statistical control frequently used in causal-comparative research is analysis of covariance. In predictive research, part and partial correlation are used. Population controls involve excluding from the study subjects that may contaminate the research. For instance, a strong relationship had been initially established between early retirement and early death. However, when people who retired early because of illness were removed from the population in subsequent research, the relationship diminished.

Four types of research have been identified corresponding to the four levels of knowledge described earlier. They are experimental research for the purpose of establishing control, causal-comparative research for the purpose of explaining cause-effect, predictive research for the purpose of predicting outcomes, and descriptive research for the purpose of describing variables and relationships among them. Several subtypes of descriptive research were also described. Each type of research suggests the general parameters of a research design, by indicating the comparisons to be made in conducting the type of research.

To complete the research design, other comparisons that are going to be made are added. Additional comparisons are often made to check on extraneous or error variables that might contaminate the research. Finally, controls are determined to reduce or remove the effects of potential extraneous variables. (Extraneous variables represent alternate hypotheses that in addition to the research hypothesis can account for the research results.) So in essence, a research design is a description of the comparisons to be made and the controls to be exercised to test an hypothesis.

The comparisons specified in a research design are often made to determine relationships between and among variables. In order to test relations, it is necessary to detect similarities and/or differences between variables. Statistical means of detecting similarities and differences are often taught in school. However, statistics are not typically taught as various means of determining relationships by detecting similarities and differences between quantitative variables. And other means of detecting similarities and differences are not routinely taught. Consequently, scholars are not often aware of the many different ways of establishing relationships.

Table 4.1 displays and encapsulates procedures for establishing relationships by detecting similarities and differences. Means of checking and controlling

Table 4.1
Procedures for Establishing Relationships by Detecting Similarities and Differences

	Similarities	Differences	Both (parts/whole schemes)
Logical Procedures	Correlation research	Experimental research	Taxonomies
	Reliability	Causal-comparative research	Class inclusion hierarchies
	Objectivity	Contrasting categories	Task analysis hierarchies
	Deriving categories		
Statistical Procedures	Pearson coefficient	Parametric techniques:	Bar graphs (histograms)
	Spearman rho	t-test	Pie charts
	Eta	ANOVA	Crossbreak tables
	Scatter plot	ANCOVA	
	Factor analysis	MANCOVA	
	Coefficient of determination	Nonparametric techniques:	
	Path analysis	Mann-Whitney U test	

112

Multiple regression

Discriminant function analysis

Reliability formulas:

    Cronback Alpha

    Kuder-Richardson

    Test-retest

    Split-half

Kruskal-Wallis

Sign test

Friedman ANOVA

Chi square

Procedures for Dealing with Extraneous Variables (Alternate Hypotheses)

Controls	Checks	Threats
Matching	Counterbalancing	Internal validity:
Covariance	Control groups	Situational factors
Partial correlation	Stratifying	Population factors
Subject replacement and oversampling (mortality)	Replications	External validity:
Keep situational factors constant and nondisruptive		Sampling bias
Keep population factors constant and appropriate		Sampling imprecision

Source: Author.

extraneous variables are added to provide a more complete picture of the factors that need to be considered in executing a research design.

The logical procedures deriving and contrasting categories as well as deriving taxonomies, class inclusion hierarchies, and task analysis hierarchies are used to derive relationships in qualitative research. And of course, statistical procedures are used to derive relationships in quantitative research. Research would benefit a great deal if both qualitative and quantitative procedures were used as needed to test hypotheses. At present, those who practice qualitative and quantitative research seem to go their separate ways as different cults, failing to avail themselves of the other's procedures.

PURSUING SOCIETAL AND PERSONAL IMPROVEMENTS

Scientific techniques can be applied within the control cycle to achieve societal improvements. Typically, societies with means subsidize research and development projects to bring about improvements of great consequence to their citizens. These are frequently large projects aimed at removing social pariahs such as poverty and dread disease. The United Nations and countries such as the United States that can afford such projects subsidize them.

Not so long ago the United States embarked on a nationwide research and development project to conquer cancer. Following the control cycle format (see Figure 4.1), *projecting improvements* was the first undertaking, which might have been projected as the eradication of cancer.

Continuing with the control cycle format, in *deriving treatments* to achieve the projected improvement, in general three major categories of treatments were derived: radiation, chemical, and surgical treatments. In a large-scale research and development project such as this, it is advantageous to test the effectiveness of a number of treatments in each category as alternate hypotheses in each successive round of testing. In this way, more comprehensive and valuable clues can be obtained for further research in the next rounds of testing.

In *implementing treatments*, treatment administrations are monitored to detect and correct deviations from specifications of the chemotherapy, radiation, and surgical treatments to ensure that treatments are actually being tested as intended. In *assessing achievement*, effects of radiation, chemical, and surgical treatments on the eradication of cancer are assessed after each round of treatment. And in *diagnosing causes*, clues are gleaned from the effects of specific radiation, chemical, and surgical treatments that were administered. Treatments for future testing are chosen based on the relative effectiveness of previous treatment applications and cogent theory. The process continues round after round as the eradication of cancer is hopefully converged on by successive approximations.

This rendition of the control cycle to societal research is suggestive of the way a large-scale project would be approached. It is not meant to be an accurate description of the way cancer research has been conducted. The point to be made is that applying the control cycle to achieve desired improvements can be much more productive than applying the traditional scientific method. Furthermore, modern research and development appears to be moving in this direction, although it may not be following the control cycle precisely as presented. The traditional scientific method may be appropriate if the aim is to advance knowledge. The control cycle appears to have great advantages if the purpose of research and development is to achieve desired improvements.

The control cycle also can be applied advantageously to the achievement of personal improvements when achievement of personal improvements requires a systematic, disciplined, protracted approach—for example, when the desired improvement is substantial weight loss. Of course, derived treatments would be weight loss regimens that might include diet and exercise. The regimens would be implemented, their effectiveness would be assessed, and if necessary, the regimens would be modified based on the diagnoses of causes of outcomes until the desired weight loss is achieved.

It is important to realize that the control cycle or any other strategy designed to improve the quality of life cannot succeed unless basic human factors are taken into account, worked with, and worked around, as need be. All too often, plans to improve the human condition fail because the nature, malleability, and intransigence of humans are not understood or dealt with realistically. Any strategy that hopes to be successful must be extended to take basic human factors into account.

The control cycle is a logical strategy for achieving improvements. However, human behavior is governed by psychological laws rather than logical laws. Two of the most potent psychological determinants of human behavior are motivation and intelligence. Motivation provides the impetus to improve. If people are resistant rather than inclined to pursue an improvement, it is much less likely that the improvement will be achieved. Intelligence provides the wherewithal to achieve a great many improvements that humans seek. If people do not have the intelligence required to achieve an improvement, it is unlikely that they will achieve it. In Chapter 5, motivational factors will be discussed. Chapter 6 takes up the intelligence issue.

Chapter 5

Working with Human Motivation

It is impossible to improve the quality of people's lives without taking into account their motivation and the means of satisfying it. Motivation manifests people's desires and determines what they will go after and what they will try to avoid. In Chapter 3, when definitions of quality of life were being considered, I said that from a personal point of view the quality of a person's life depends primarily on his or her ability to pursue personal aspirations. Well, personal aspirations are pursuits in which people are motivated to engage. To help people fulfill their aspirations, we need to more thoroughly understand their motivation. And although to some extent motivation varies from person to person, to help improve the quality of human life in general, it would be beneficial to identify prevalent motivation that impels people's actions most of the time. Identifying prevalent human motivation that can be enlisted and worked with to improve the quality of life is the mission of this chapter.

Huge sums of money are spent on motivational research. Business and industries want to know how to entice people to buy their goods and services and how to encourage employees to be more productive. Welfare agencies want to determine how to get people on welfare to want to work. Charities want to know how to entice people to donate more money. Crime prevention agencies would like to know how to get convicts to choose to obey the law. And parents want to know how to induce their children to cooperate without exerting undue coercion. Little can be done to help people fulfill their aspirations or to provide incentives so that they will do voluntarily what it is necessary for them to do to be productive members of society without understanding motivation.

Although knowledge of human motivation is still quite primitive, enough is known to help individuals and societies work to improve the quality of life. However, the goal is not easy to achieve. Many motives are difficult to work with. Many are transient. They are here one minute and gone the next—and hence unreliable. For example, anger is frequently fleeting. One minute a parent may be angry with his child, and the next minute be forgiving. Further, many motives cannot be satisfied in most situations without violating the law and undermining the fabric of society. For instance, there are usually laws that restrict the free expression of the sex motive. And the unbridled satisfaction of the sex motive can interfere with family cohesiveness and productive work. Consequently, sexual fidelity is encouraged by law, and sex is not allowed in the workplace. Finding a motive that is reliably present and that can be enlisted to improve the quality of life has been very difficult until recently.

As complex as human motivation may be, one motive has been identified that is reliably present and can be worked with to improve the quality of life: the control motive. The remainder of the chapter is devoted to explaining how and why the control motive can be enlisted to improve the quality of life.

Most people have some idea of what *control* means. They probably know that exterminators control bugs, pilots control airplanes, parents try to control their children, lunatics are out of control, polio is under control, and we all need to control our weight and our expenses. Control is so essential to life that people can hardly survive without learning something about it. If they have occasion to consult the dictionary, they might find that control can mean "to regulate or direct," and that it can have other meanings and shades of meaning. For our purposes *control is defined as influencing things or people, including oneself, to bring about desired outcomes.*

With this in mind, there is a need at the outset to review the evidence demonstrating that people are motivated to control and that control is important to quality of life. Many people are more familiar with other motives such as envy, guilt, hostility, sex, hunger, and love than they are with the control motive.

There can be little doubt that feelings of control are a major factor in improving health. People who believe that they have control over outcomes in their life fare much better healthwise (O'Leary, 1985). Nursing home studies show that inducing feelings of control in older patients improves their well-being, including a reduction in both disease and mortality rate (Langer and Rodin, 1976; Schultz and Hoyer, 1976; Langer et al., 1979; Banzinger and Roush, 1983). So people who feel that they have control over their life enjoy better health. And when people who have lost control are made to feel more in control, their health improves. Furthermore, people who feel more in control take better care of themselves and greater responsibility for their health needs (Wallston et al. 1976). They learn more about their illnesses (Toner and Manuck 1979), and they are more likely to benefit from health education programs (Waller and Bates 1992). They also are less withdrawn, have more active social lives, and feel better about their lives and environments (Lemke and Moos 1981; Moos 1981; Moos and Ingra 1980; Hickson, Housely, and Boyle 1988). The evidence becomes more compelling as research accumulates. Gaining control can have curative benefits.

Furthermore, Donovan and O'Leary (1983) concluded from their research as well as from their review of the research of others that alcoholics appear to experience less control over both interpersonal and intrapersonal sources of stress in their life than do nonalcoholics. Doherty (1983), citing his own research and the research of others, concluded that people who perceive themselves as in control of outcomes experience greater marital satisfaction. Finley and Cooper (1983), after a review of 98 research studies, concluded that perceived control of outcomes is associated with greater academic achievement. Gordon (1977) confirmed the association between greater perceived personal control and academic achievement and found an association between greater perceived personal control and higher self-esteem as well. Pervin (1963) showed that when confronted with threat, people prefer and suffer less when they are able to predict and control the threatening event.

People who have a sense of control over their lives fare much better than those who don't. A key factor in improving people's well-being is instilling in them a sense of control and helping them gain control over events in their lives. As they feel a greater sense of control, they exercise more control and become more content, active, and effectual.

People attempt to control almost anything that worries them, for example: smoking (Kuchkremer, Milnneker, and Block 1991), drinking (McMurran 1991), drug use (Carroll, Rounsaville, and Keller 1991), compulsive gambling (Lesieur and Rosenthal 1991), and diet and weight (Foreyt and Goodrick 1991). Educators try to control aggressive students (Etscheidt 1991); some abusive parents try to control their destructive behavior toward their children (Acton and During 1990); and psychotherapists help phobic children control their fears (Ollendick, Hagopian, and Huntzinger 1991).

Now that the importance of the control motive to quality of life has been established, control needs to be understood better. To get what they want, people need to influence both people and objects in their environments. And to control their environments, they need to be able to control themselves. Self-control is prerequisite to environmental control. If people can't control themselves, they can't control anything else. To take pills, they must be able to open the containers and put the correct number in their mouths and swallow them. To get along with others, they must control their tempers and say things that are ingratiating to them. Further, self-control is necessary to achieve personal as well as environmental goals. It is necessary to lose weight, build body strength, control drug intake, refrain from sin, and take time out to rest when weary.

Objects are usually controlled by physically manipulating them. Turning on a stove, closing a door, and driving a car are examples. Sometimes people are controlled in the same way. Infants are manipulated when they are diapered, as are old people when they are restrained in their beds. However, when people can understand what they are told, attempts to control them are usually linguistic. Subordinates are given orders. Drill sergeants give marching orders. Doctors give orders to nurses and to their patients. People receive orders in writing when they are billed to pay taxes or are summoned to court. On the other hand, when people are under no obligation to take orders, attempts to control them are through requests and persuasion. Sometimes it's only necessary for people to ask for what they want, be it to get advice from a lawyer or to borrow a dime. Persuasion is needed when there is a need to compel someone to grant a request—for example, when a person is attempting to talk a policeman out of a traffic ticket or to convince a judge of his innocence or a caretaker of his ability to care for himself.

Cooperation is often needed to control outcomes. Members of sports teams cooperate to win games; work teams, to manufacture products; and children, to help aging parents. Team members must subordinate personal goals to group goals if the group is to succeed. Individuals might be benched, fined, fired, or divorced if they sacrifice the common good to indulge their personal desires. In other social situations, competition is required to control outcomes, for instance, when people try out for a sports team, apply for a job, try to win someone's affection, or try to beat an opponent in a game.

There are many ways to try to control others. Bullying, blaming, belittling, dictating, preaching, and persuading are among the more assertive ways. Pleading, flattering, coaxing, seducing, and bestowing favors and gifts are among the more unintrusive ways. Unintrusive attempts to control others are more tactful and may be required if one is a subordinate. Superordinates often can adopt more assertive modes of control without suffering the recriminations that subordinates can suffer. *Domination* and *control* are not synonymous. It is quite possible to control people without dominating them.

The importance of control to humans is manifested by its frequent inclusion in theoretical treatises that attempt to explain human behavior. Rotter's (1966, 1975) locus of control theory is quite well known. It contrasts the behavior of people who believe that control of events is internal with people who believe that control of events is external to themselves. Gibbs (1989) describes control as sociology's central notion. In prediction theory, control is the most prevalent human motive (Friedman and Lackey 1991). Glasser (1984) views people's actions as attempts to control their lives. Bales (1950) explains group behavior as attempts to control interpersonal relations. DeCharms (1968) asserts that humans are motivated primarily to control their environments, while Langer (1983) states that choice is directed toward controlling goal achievement. Kelly (1963) describes humans as prototype scientists seeking to predict and thus control the course of events. Schulz and Heckhausen (1996) use control as the motivational force underlying their life span model of successful aging. They contend that the motivation for primary control fuels development throughout the life course—our model indicates that person-environment interactions are driven by the motivation for primary control. They claim that extensive empirical research supports their contention. There are many theorists that view control as a major force motivating human behavior.

Now that the importance, nature, and aspects of control have been clarified, we are ready to consider how motivation to control functions, first, from a personal perspective and, then, from a societal perspective.

PERSONAL IMPLICATIONS OF THE CONTROL MOTIVE

The answer to the question, Why do people behave the way they do? is, Most of the time they behave to improve their control of outcomes. This does not deny the presence of other motives. It simply indicates that of all human motives motivation to *improve control* of outcomes is the most prevalent.

It may be difficult at first to understand that motivation to control is so prevalent, because people tend to be more aware of more intense and fervent motives, such as hunger, thirst, physical safety, and the desire for sex. However, on closer inspection, some of the more intense and conspicuous motives may not be as prevalent in humans as they are in lower creatures. There is no denying that humans are interested in survival and personal safety, like other animals, but humans are willing to sacrifice their lives for their ideals. The human suicide rate is a serious problem, and with the advent of intricate, effective life-support systems, euthanasia is becoming more common. Yes, people are interested in procreating like other creatures, but they plan parenthood. Some are not willing to make the personal sacrifices it takes to raise children, and some seek abortions. Sure, people want to satisfy their appetites, as other animals do. However, some deny themselves food to diet and to fast on religious holidays, and some deny themselves sex to avoid disease and to fulfill sacred matrimonial and religious vows. At second glance, there are significant disparities between human motivation and the motives of other animals.

Although motivation to improve control is most prevalent, sometimes its presence is obvious, sometimes not. At times, we are keenly aware of our motivation to improve our control—when we are persuading a rebellious child to obey, when we are attempting to entice someone to like or love us; and when we are trying to force trash into an overcrowded trash can. At other times, motivation to improve control is masked by more intense motives, and this can keep us from realizing that it is so commonplace.

For example, when people are hungry, they crave to satisfy their hunger; when people are sexually aroused, they feel their passion; and when teenagers want to learn to drive a car, they make it evident that they are vehemently motivated to drive. On the other hand, it is not so obvious that motivation to control is an underlying motive in all these cases. Yet it lurks behind the scenes, pressing persistently for control of the sources of satisfaction—the acquisition of food to satisfy hunger, the acquisition and use of a sex partner to satisfy the sex drive, and the acquisition of a driving instructor to satisfy the desire to drive. A retired corporate chief executive officer who is dissatisfied with relaxation, recreation, and idleness may not realize that what he or she misses most is controlling people and resources to make a profit. A person with new dentures is initially more aware of pain or discomfort than of the need to control them to eat.

It is only when the imbroglio of human motives is penetrated and sorted out that the control motive can be seen as such a prevalent, generalizable motive underlying, energizing, and giving direction to such a vast variety of pursuits. Looking at individuals as individuals, they can be seen at one time attempting to satisfy their hunger, at another time, their sex drive, at another time, their desire for wealth, and so on. Underlying and common to each of the specific pursuits is motivation to improve control of the sources of satisfaction so that they can satisfy themselves at present and when the specific desire emerges again at future times. Looking across people, we may observe one person seeking to satisfy his hunger, another his sex drive, another his desire for wealth, and so forth. We find that underlying and common to each of these pursuits is the motivation to improve control of the sources of satisfaction now and in the future. So the conclusion that the motivation to improve control is a most general motive emerges compellingly from probes to discover what energizes and guides human behavior, whether we observe personal or interpersonal behavior

Once people are alerted and look for the prevalence of the control motive, they can find all around evidence of people's preoccupation with control. People talk about controlling their cars, their board meetings, their employees, their relatives, their incomes, their weights, the possibility of war, the threat of natural disasters, and their professional reputations. And they control their property from beyond the grave through the use of wills and trusts. *The Psychology of Control and Aging* (Baltes and Baltes 1986) provides some penetrating insights on control and aging.

People seek help when they can't control things on their own. Children depend on adults to control things for them. And grown-ups seek help from trained professionals when they lack expertise. Doctors, dentists, lawyers, mechanics, hair stylists, and plumbers frequently provide assistance. Some turn for help to people imputed to have occult talents—medicine men, rainmakers, mediums, exorcists, clairvoyants, magicians, and the like. In olden times, it was said that the master magician "cohabited with the elements, vanquished nature, mounted higher than the heavens, elevating himself to the archetype itself with whom he becomes cooperator and can do all things" (*Encyclopedia of Magic and Superstition* 1974, 12). Still others practice rituals to gain control. They pray. They practice such superstitious behavior as refusing to walk under a ladder. They wear lucky charms such as a rabbit's foot and lucky articles of clothing. Teams adopt mascots. Some people use image dolls to injure, kill, or seduce others. In short, when people cannot control outcomes themselves, they go to great lengths to gain control, some of which are mysterious and bizarre.

Even when help isn't needed, people like to get others to control outcomes for them to conserve energy and to make them feel more secure. It is comforting to feel that others are there to do your bidding when you want them to. Erich Fromm in his book *Escape from Freedom* (1971) explains the tendency of many people to relinquish freedom in order to be taken care of. They want to control outcomes as much as anyone, but like children, they want to control by wishing and requesting. He explains why this inclination to preserve dependency and avoid the responsibility and burden of autonomy allows dictators and "Big Brothers" to rise to power. Surrendering to others the power to control things for the individual often backfires. Dependency slows maturation, and dependents most often become disillusioned because their benefactors inevitably fail to grant enough of their wishes. All people want to control. Most often autonomous, self-reliant control turns out to be more reliable than dependent control, although people often wish it were otherwise. Many competent old people coaxed to give up control of their assets to their families who promised to take care of them wish they hadn't.

People not only want to control to get things: they find improving their control satisfying in itself. They feel good about themselves when they are a take-control person, in control of their lives. They feel more competent, more capable. Every time they set a goal and control its achievement, they are self-satisfied.

LOSS OF CONTROL: THE BANE OF DISABILITY

People past their prime probably are aware of how depressing physical losses can be. However healthy, active, and productive they might be, they can't help noticing that they do not have the strength, endurance, and agility they once had. They may take good care of themselves, eat healthy food, exercise regularly, and be able to brag about how healthy they are for their age, but they most probably notice some fading in their performance. They might not be able to run as fast or as far as before, they may no longer score as low when they play golf or be seeded as high in tennis competition, or they may not be able to lift as much weight as before. Whether they discuss it openly or prefer to avoid being reminded of it, in all likelihood, they have some inkling of their diminished physical capabilities.

Others have described the plights of disability far better than I can. What I am stressing now is that of all the losses people suffer because of disability, loss of control is psychologically the most devastating. As upsetting as physical, social, and financial losses may be, and as much as these losses may contribute to loss of control, it is loss of control that is most psychologically damaging, causes the deepest depression, and destroys the will to live. This, in turn, accelerates deterioration and death (Thomas et al. 1992). To elaborate, loss of leg strength and endurance undeniably causes stress. Much greater mental stress is caused when these physical losses result in loss of control of outcomes, for instance, when people can no longer go after, make, and get the things they need. Still, their outlook and zest for life will improve when they regain their ability to fend for themselves. To derive the satisfaction that regained control brings, they need not regain the leg strength and endurance that was lost. They need to regain the control of outcomes they have lost. This might be attained by the use of a cane, a walker, a wheelchair, and/or a motorized scooter or chair.

Now, it is important to understand that people's attitudes toward life will improve whenever their control of outcomes improves, whatever the circumstances may be. Their attitudes will improve when they regain control of something they previously lost control of. In general, loss of control will dampen people's spirits and improved control will be satisfying. Improved control is satisfying even if there are annoying concomitants. For instance, parents may not enjoy punishing their children, but they do it when they believe it will make them cooperate.

Victims of loss of control usually identify their problems as some other, more tangible loss, without necessarily realizing that loss of control is an underlying cause. When they break a leg, their attention is drawn to the pain in their leg and the cumbersome cast they must wear for a while. They don't lose control to any great extent because they are provided with crutches or a walking cast and may still be able to drive a car. In a while, they adjust to the cast and consider the main setback to be the things they have lost control of, perhaps scratching their leg, bathing as usual, or the sports they must give up for the time being. But they are not necessarily aware that they are worried because of their loss of control. Loss of control is not as intrusive as a broken leg.

One symptom that makes loss of control apparent is increased dependency. Loss of memory requires people to become more dependent on writing things down, making "get and do" lists and entries in appointment books. Loss of control of movement requires them to get others to go shopping for them and to bring things to them at home.

The more people lose control, the less interest they have in prolonging their lives. Living wills are made because people realize beforehand that they do not want to continue to live with no conscious, purposeful control of their lives. Purposeful control is gone when people are unconscious and need life-support systems to live.

Loss of control can be mind-boggling and unbearable. When the last vestiges of control seem to be slipping away from people and they feel they can't get a handle on their lives, the need to end the intolerable quandary can become a preoccupation, and any way out can be a preferable alternative.

Many forms of mental illness may reflect the control-oriented nature of the individual being thwarted in some fundamental way. Often, people speak of adjustment problems as manifesting themselves in the individual withdrawing from reality. Can it be that they are withdrawing from a social or physical environment over which they have lost control? Could it be a little like the young child who won't play because he cannot have his own way—control?

Clearly, some of our virulent emotions seem to be triggered by loss or failure to control: insecurity, anger, jealousy, envy, frustration, panic, and so on. It may even be argued that the most emotional moments in people's lives are those when they are threatened by events over which they have no hope of control.

It's important that professionals as well as the public be sensitive to particular losses that affect people's survival and the quality of their lives. A special effort

Contral of Intake

Control of intake is necessary for survival. The intake of oxygen and food keeps people alive, although the intake of oxygen is, for the most part, controlled autonomically. Breathing and aerobic exercises as well as refraining from smoking help maintain healthy lungs. And when breathing is at risk, medications such as bronchial dilators and anti-inflammatory drugs can be used to improve lung function.

Eating is a different matter. Sucking is autonomic, but unlike inhaling, sucking does not automatically bring life-giving substance into the body. In infancy, nourishment can be taken in only if adults provide it. People do not gain complete control of the intake of nourishment until they can provide it for themselves and are able to put it in their mouths, swallow it, and digest it, all of which can become problematical with disability. A diminishing sense of taste, indigestion, depression, mouth sores, dentures, gum disease, and tooth decay can dissuade people from eating. And when they eat, they may choose food because it is not irritating rather than because it is healthful.

The intake of drugs is another serious matter. "Approximately 34 percent of older adults take three or more prescription medications. It is not uncommon to encounter older adults with multiple chronic conditions taking eight or more medications simultaneously" (*Vitality For Life* 1993). Unfortunately, many of the elderly suffer from memory loss and can't be relied on to take the medicine they need on schedule in the dosages prescribed. Further, the side effects of some drugs impair mental function. It is not uncommon for the sedative effects of some medications to cause people to sleep through scheduled medication times.

It's not just the intake of prescribed drugs that must be monitored carefully; it's the intake of alcohol and other temporary mood elevators that can imperil life. More and more people are tempted to take drugs to escape boredom, confusion, and demands they are unable to meet. The point is that for people to take control of their lives, they must take the responsibility for getting the advice they need from doctors, nutritionists, and authoritative publications, taking in the nutrition they need, and rejecting drugs and other substances that can be harmful. Of course, drinking alcohol in moderation can be beneficial, with the realization that the elderly can have more bizarre, prolonged, and extreme reactions to lesser amounts of drugs. Because depression is a malady that saps the will to live and thwarts people's interest in providing for their own welfare, victims are likely to succumb to drugs and lose interest in eating. To avoid the ravages of depression, people must remain actively engaged in controlling their own lives.

Intake through the senses is seldom a matter of survival, but it can make life more enjoyable or more annoying. Sexual pleasure is gained through touching, kissing, and intercourse. Pain is felt when passion is unrequited. Intake brings aesthetic pleasure when beauty is beheld, and pain when grotesqueness is experienced. Love and affection are initiated through the senses when the presence of someone gives pleasure, and deprivation of a loved one causes pain.

All that people learn requires intake through the senses. Learning takes place casually, by happenstance, as people record in memory the experiences they come upon. Learning takes place systematically, on purpose, when people are subjected to goal-directed instruction by parents and other teachers. Learning by reading involves taking in through the eyes, while learning by listening requires taking in through the ears. Learning is enriched and more complete when more of the senses are involved.

Intake through the senses is especially important to people who rely mainly on passive forms of recreation. Television, movies, the theater, record players, and books provide pleasure without strain. And given the opportunity, people are usually thirsty to learn. More and more opportunities are offered to travel, take courses, and earn degrees. People are not interested in learning for the sake of killing time, they are interested in increasing their knowledge as a means of improving their control. When old ways of doing things no longer work as well, they are interested in learning new ways to achieve their ends. When improved control can be achieved, one can teach an old dog new tricks.

Intake has significant psychological as well as physical consequences. When control of intake progresses pleasantly without trauma, people learn to trust the world they live in. They are getting from their environment what they need and want. They gain faith in life and what the future may hold for them. They are able to enjoy intimacy and express affection and love for others. They have learned to accommodate and regulate their behavior to comply with external demands.

On the other hand, if the taking-in process is unsuccessful and painful, people mistrust their environments. They feel forsaken by their providers. They feel empty, deprived, wanting, and in peril, desperate to survive. As a result, they become grasping, envious, and jealous and regard the environment as punitive and destructive. In general, control of intake engenders trust; inability to control intake engenders mistrust.

People who have been trusting can become distrustful if they lose control of intake and the things they need, want, and are accustomed to are not forthcoming. It's important to realize that trust is based not so much on what people are able to control themselves but on feeling that they will be taking in what they want and need. To be trustful when they are no longer able to provide their own needs, they must be able to rely on providers. For the disabled to remain trusting, their caregivers must be dependable in providing their needs.

Control of Movement

Control of movement is another vital control function that can diminish with disability. Without being able to move, people are unable to go after the things they need and want, and they can't avoid threats to ensure their own safety. Control of movement begins in the crib as infants reach for objects and draw them near, usually to their mouths. It culminates when adults are able to control their bodies at will to walk, run, jump, climb, kick, stoop, squat, punch, push, pull, lift, reach, write, ride a bicycle, and drive a car. In addition, they are able to use public transportation to take them where they want to go. If they are fortunate, they own their own car and possibly a boat or plane. Loss of physical movement seriously reduces people's ability to fend for themselves.

People learn to control social as well as physical mobility. Children learn that being promoted to the next grade is desirable and rewarded and that they are expected to progress from elementary school to high school to college. When people enter the world of work, they learn about being promoted in the organization and earning higher status and more money. They learn about the pyramidal hierarchies used to define business and government organizations. It is possible for them to move all the way to the top to become president of a corporation or even the nation. Social mobility is almost as important as physical mobility. To succeed socially, people must learn to control their exchanges with others to their advantage. Shrewd exchanges result in financial profit and gains in status. Loss of money and status reduces control. Loss of income reduces purchasing power, and loss of status reduces the realm of authority over others.

In general, movement enables goal achievement, going after the things people want and avoiding threats. Goal achievement may involve seeking a promotion or evading the boss's wrath, getting a second medical opinion, or avoiding fattening food. It always involves behavior, perhaps something as innocuous as applying to be accepted in a retirement village or something as violent as killing the enemy in wartime. Whatever goal people may wish to achieve, goal achievement always involves purposefully pursuing it.

When people's movements achieve their goals, they feel effective and gain confidence that they will be able to achieve their objectives whether a little assertiveness or an attack is required, whether they simply need to avoid someone's presence or they need to flee. So the successful control of movement instills in people the self-assuredness of knowing that they are able to purposefully take the actions required to deal with their environment effectively. Control of movement clearly instills more than control of intake. Control of intake engenders the feeling that we will be taken care of. Control of movement engenders the feeling that we can fend for ourselves.

Conversely, when control of movement is unsuccessful, we feel inept, unable to take care of ourselves. We feel that our actions are to little avail; our efforts get us nowhere. We feel inadequate. In the extreme, we feel helpless.

Loss of control of movement is devastating to the disabled. When people become unable to drive a car, they are far less active in community life and spend much more time at home. This interferes with shopping; visiting friends; attending meetings, movies, and luncheons; and going to work or school. To many, a car is regarded as a necessity. Some rent cars so that they do not have to be frustrated and inconvenienced when their cars are being repaired. It can be devastating when people can no longer drive.

Loss of control of body movements is even more devastating because these losses are signs of feebleness, signs of regression toward infantile helplessness, signs that people are no longer able to take care of themselves. Loss of control of walking threatens people's ability to move around the premises to do things for themselves and to move out of harm's way.

Social mobility, too, is compromised with disability. Sometimes loss of bodily movement impairs social mobility. People who cannot get around well are often not as able to assume the responsibilities of work and community leadership as well. Loss of memory imposes a serious impediment to both social and work aspirations.

Control of movement is fundamental to life satisfaction, both physical and social movement. Loss of control of movement is inevitable and crushing to the spirit. It results in helplessness.

Control of Production

Control of production is a third vital control function. It differs from control of movement, which enables people to go after the things they want. Control of production enables people to *make* the things they want. Control of production begins with babies' control of the production of their feces and urine. Their production of feces in the potty or toilet is noticeable to them. They like to look at the stool they produced and show signs of pride and pleasure at their accomplishment. The control of the production of excrement is the child's first autonomous act of control, unlike early feeding, which is controlled by an adult provider. Since people are motivated to control, it seems quite natural that exercising control would be pleasurable and that exercising control autonomously would inspire pride. Producing things on one's own does require autonomy, and it appears to engender pride, because the producer can take primary credit for the achievement. The producer made it and brought it into the world all by himself or herself. This pride seems to be associated with the successful production of anything.

Demands are made of people to be productive. In school, youth are expected to produce good grades, and at home, they may learn about cleaning, gardening, and cooking. In maturing, they should progress from producing orgasms by themselves through masturbation to producing orgasms with a sex partner. When they are adults, hopefully they will become concerned with producing mutual orgasms, with producing goods and services to earn a living, and with mating and producing offspring.

Control of production can entail creativity. Control of intake enables people to receive enjoyment from art and music. Control of production enables them to produce art and music. In addition, people produce ideas. Some are fantasies generated in night dreams or daydreams. Others are plans for achieving goals. Humans are capable of producing innovations, first producing plans of things that have never been seen before, then carrying out the plans to materialize the innovations. The production of innovations is a hallmark of human superiority over other creatures and is largely responsible for improving the quality of life from one generation to the next.

Individuals' successful development of the ability to control production engenders autonomy and a sense of power in knowing that they have control of their own destinies. They can transform circumstances to make them turn out as they wish. They can produce many of the things they need and no longer need to depend on others to provide them. Nor do they need to chase after as many of the things they need. Human progress from a hunting to a farming culture enabled people to produce food in their own backyards so that they did not have to hunt for four-footed protein and gather other edibles. The ability to produce what they need so that it is available when they need it is the epitome of control. It is far better than relying on others to provide it or to try to find it when they need it. Control of production brings a sense of pride and dignity as well as power. People are proud of their ability to produce their needs and feel dignity because they do not need to beseech others to provide their needs.

The failure to develop control of production prolongs dependency and engenders self-doubt and feelings of inferiority and shame. People are acutely aware of the productivity of others. Those who cannot produce mutual orgasms with a sex partner, offspring, or a livelihood are destined to consider themselves inferior.

Loss of control of production need not be immediately life threatening, but it certainly affects the will to live. It undermines pride and dignity and results in feelings of shame and inferiority.

Control of Decision Making

Control of decision making is a fourth vital control function. Probably the most damaging loss *is* loss of control of decision making. Slight loss of memory does not render decision making ineffective. Although it may be embarrassing to forget a name or an important date, forgetting information needed for decision making merely requires people to get the information again. It is inconvenient and time-consuming but not incapacitating. Severe loss of memory is another story. People lose orientation that is central to decision making. They can't make adaptive decisions if they don't remember who they are, where they are, where they are coming from, or where they are going.

For humans, a most critical loss is the ability to arrive at sound, well-considered decisions. Loss of physical abilities merely reduces the options people can consider in making a decision. Loss of mental faculties can preclude decision making, which for the most part precludes purposeful control. It is not that forgetful people can no longer produce ideas; it's that the ideas they produce no longer enable them to make effective decisions. Many become preoccupied with "the good old days" when they were in control of their lives. But, sadly, they are unable to produce memories that allow them to use their past experiences to solve their present problems. This makes it quite difficult for them to make the decisions they are called on to make daily. Control is subverted when people are unable to use their experiences to make decisions. On the bright side, I am reminded of a 70-year-old comedian who jokes about the benefits of his forgetfulness. He says he likes meeting so many new people every day. He can hide his own Easter eggs and enjoys reruns as much as movies he sees for the first time.

Decision making is the key to control. The more people decide their own activities, the more they are in control of their lives. They may not have the physical ability to carry out their decisions. Still, if they make their own decisions, they maintain control, even if others help them carry out their decisions.

Conversely, they can't maintain control of their lives if they don't decide their activities, no matter how much physical strength, endurance, and agility they may have. They may be able to drive a car, cook, hold a job, or compete in sports, yet to the extent that others dictate what they do, when they do it, and where they do it, they are not in control of their own lives.

One of the mistakes people make is to believe that because they have suffered a physical loss, they necessarily must lose control. Being concerned about physical losses is natural, but prolonged distress over those losses is depressing and can be debilitating. When people begin to slide into the doldrums of despair because of a physical loss, they need to be reminded that they need not lose control as a result. That is the time to start thinking of ways to compensate for their losses and get on with their lives. Only if they continue to brood and long for their more physically adept years are they certain to lose control.

Brains are a far greater asset than brawn. There are greater rewards for the decision makers than for people who carry out the decisions. Executives make decisions; their employees execute them. It's wise to keep in mind that it is humans' big brains that enables them to control the planet. Animals that are physically much bigger, stronger, faster, and more agile than people do our bidding. We kill them at our will for food and at our whim during hunting season. A human with an intelligence quotient (IQ) of 80 is an Einstein compared to the most intelligent ape. And we used our intelligence to invent tools and machines that with our manipulation are much more capable of doing menial tasks than we are. Cranes lift weights better than we can. Robots weld seams better than we can. And thermostats control temperatures better than we can. We make the decisions; they carry them out.

The reason why loss of control is so debilitating has been discussed in general and shown to be incapacitating in four vital areas: intake, movement, production, and decision making. An important thing to consider is that in whatever area loss of control might occur, it is psychologically distressing, and if pronounced and prolonged, loss of control will result in depression. Conversely, dejection can be relieved by increasing a person's control in any area, because people find improved control satisfying and uplifting. A study reported in *Health after 50* found that when frail people became dejected, their outlook and satisfaction with life increased by engaging them in activities that improved their control. Activities were as simple and mundane as engaging them in caring for pets, growing potted plants, feeding birds, and helping others with their chores (Johns Hopkins 1994). These are treatments for despondency prescribed by psychotherapists.

SOCIETAL IMPLICATIONS OF THE CONTROL MOTIVE

In their efforts to improve the quality of life of people, societies can be more successful by inducing their citizens to obey the law voluntarily and contribute to the achievement of societal goals than by forcing them to comply and punishing them for noncompliance. Law enforcement and prisons are exceedingly expensive, people resent being coerced, and societal disintegration occurs when core values are excessively ignored and protested. To get people to do voluntarily what it is desirable for them to do requires, first, enough knowledge of human motivation to be able to provide incentives to gain their cooperation. For would-be incentives to work, they must satisfy basic motivation. Lowering the insurance rate of drivers who don't get traffic tickets is supposed to reduce accidents. But does it? Will drivers who have frequent accidents stop having accidents to avail themselves of lower insurance rates? Quite often, would-be incentives don't work because they do not satisfy motives as expected. Second, getting voluntary cooperation requires an understanding of constraints that must be dealt with in order to succeed, as illustrated in the adage, Don't try to teach a pig to sing. You won't have much success, and it will annoy the pig.

Although there are a number of motives societies might attempt to work with, the expression of many motives needs to be restrained to avoid violations of the law and subversion of the achievement of societal goals. For example, hostility needs to be restrained to prevent injury and destruction. Furthermore, unbridled satisfaction of many motives such as lust can interfere with improving the quality of life, unless one believes that the quality of life in Sodom and Gomorrah was exemplary. In contrast, the control motive can legally be enlisted and satisfied to improve the quality of life in societies.

There is no fundamental conflict between people's desire to control and social welfare. To be productive citizens, people must be able to exercise sufficient control to take care of themselves, or they will be social wards. And hopefully, their ability to control extends further, to the point that they are able to work with others to make social contributions. Society not only wants people out of institutions for the inept; it wants people off welfare, gainfully employed in a work enterprise. And control is essential to successful work.

Industrial and organizational psychologists have been striving to improve our understanding of work motivation since psychologists began to focus on, and specialize in, work as a discipline. To fathom work motivation, we must attempt to understand why people will work in one setting and not another, and why money is not a sufficient lure to ensure the desire to work. There is work that people will do without pay, and work people will not do for any amount of money. As difficult as it may be to understand work motivation, the payoff would be substantial. If work motivation can be better understood, then presumably more effective provisions can be made for job enrichment, work incentives, increased productivity, job satisfaction, and the reduction of absenteeism and tardiness.

Although from time to time other motivations will emerge, most often workers will be motivated to improve their control. What's more, workers' motivation to control can be tapped and exploited in the workplace because most often achievement of work objectives requires control. That is, control is necessary to the production of goods and services. In short, achievement of control is intrinsic to work, and workers will derive satisfaction from exercising control at work. So workers and the organizations they work for may be in conflict about many things, but both are in favor of workers exercising control of outcomes to produce the goods and services that are of mutual benefit.

To solve work problems, it is necessary to understand basic motivation and to bring the satisfaction of basic motivation in tune with achieving work objectives. Satisfying workers' appetitive motives, such as the desire for love or sex, on the job is difficult and counterproductive, if not totally destructive of the work enterprise. To maintain work efficiency, workers are required to satisfy their appetites away from the workplace. But what workers can be encouraged to do to derive satisfaction on the job is to exercise control of outcomes to produce the goods and services necessary for the success of the organization.

Extrinsic rewards do not need to be contrived if workers' intrinsic desires for control are satisfied on the job. Management would not need to spend so much time thinking of incentive plans, bonuses, perks, prizes, and fringe benefits to get workers to be productive. Workers would be deriving satisfaction from exercising control, and there would be less need to bribe them to work. When extrinsic rewards are ill conceived or overdone, they tend to detract from the importance of finding satisfaction in work itself. Since so many of us spend so many hours at work, it is a pity when we do not realize the gratification that can be derived directly from our efforts.

How is work motivation enlisted productively? By making certain that people see the contribution of their work to the improved control of outcomes. The implication is clear: When people understand that the work they do contributes to their control of outcomes, they will be more satisfied with their work.

Now, it cannot be assumed that workers' motivation to control necessarily motivates them to work with others. If they can see that working with others improves their control, they will be so inclined; otherwise, they will not. Workers will balk at working with colleagues who they predict will impair their control, and they will want to work with colleagues when they predict their control will be enhanced. (Children, even in the primary grades, will not choose others to "be on their team" for work or play unless they offer enhancement toward success in the project or game.) Moreover, workers will predict that their control will be enhanced for one of two reasons. They may predict that working together with a colleague will improve their control. In this case, they see the value of division of labor or cooperation. They most certainly will predict that being given authority over other workers will improve their control because it, in fact, increases their scope of control.

It follows, then, that an incentive that can be provided for workers is the opportunity to rise to positions of higher authority. It also follows that incentive is provided to work with others when it can be seen that improved personal control accrues because of cooperation.

Providing incentives extends to the personnel development policies and procedures of an organization. Since workers are motivated to advance to positions of higher status to improve their control, to keep morale high, organizations must provide for the advancement of their workers, and advancements must provide for increased control. The bureaucratic ruse of promoting undesirables to positions where they are out of the mainstream and ineffectual is counterproductive. Promotions should be substantive, not titular. They should reflect the importance of the workers' contributions to productivity and should not be a form of exile or a substitute for merit raises in pay. Organizations must also facilitate workers' advancement. This means that they show preference for advancement within the organization rather than the hiring of new personnel. In order to facilitate internal advancement and to ensure that new job holders are well qualified, organizations must provide for job training. This might range from paying workers' tuition to attend school to providing for on-the-job training.

Finally, there are implications for providing incentives to people the organization is attempting to hire. Prospects can be lured to accept job offers if they are able to see that by taking the positions they will be able to improve their control of outcomes. Acceptance rates can be expected to increase if prospects are convinced that taking the jobs increases their control at work and in their personal lives. In most cases, this would be a lot more compelling than attempting to entice prospects with recreational opportunities provided by the organization, the company spirit, trophies awarded to outstanding workers, liberal allowances for absences and vacations, piped music, or cheerful and tasteful decor.

Never before has research and development been as important to business and industry as it is today, and its importance continues to grow with each passing year. Accelerating change is a fact of life in modern societies. As expanding technology feeds upon itself, the public must adjust to the rapidly increasing spate of novelty and change. Businesses that do not improve their productivity continually are soon left by the wayside. It has never been more essential that businesses reinvest their profits in research and development to ensure their own survival and success.

To be effective, research and development needs to take into account people's motivation to improve their control. If the research and development of a work enterprise results in the provision of goods or services in a way that better satisfies people's motivation to improve their control, there will be a demand for it. The emphasis is on satisfying people's motivation rather than their needs. This becomes clear when we realize that people buy television sets before they provide sufficient nutrition and sanitation for themselves and their families. Promotions that appeal to people's motivation to control will tend to be more successful than promotions that do not. Successful perfume ads not only suggest that one's sex drive will be satisfied when a perfume is used: they suggest improved control over the opposite sex.

Practitioners striving to improve the quality of people's lives can work advantageously with the control motive. They can satisfy the control motive by controlling their treatment to benefit the people they treat. This approach, in turn, instills professional pride and praise from their clients. Practitioners frequently need to exercise control over their clients to get them to cooperate, which can be achieved by exploiting the control motive. For instance, dentists can suggest that their patients raise their hands to stop treatment when they are in pain. On the other hand, motives such as hostility can subvert professional practice, if expressed on the job.

People create social institutions for purposes of control. The state provides to a greater or lesser extent control over the physical resources and the social activities of the individual. Religious organizations generally serve the same control function. Schools prepare individuals for control in the future, either as individuals or as members of the society. One social institution battles another as both attempt to exercise control over the activities of the larger group. The goals of social institutions differ from culture to culture, but the purposes of the institutions are always the same: the control of the social and physical environments of the culture. Such institutions may be understood as being effective to the extent that they are successful in their control and as being oppressive to the extent that the individual is limited by them in the control of his or her personal and immediate environment. Whether the state emphasizes the achievement of group goals or individual goals is really only a matter of who controls what. That human beings, individually or in groups, will act to control is the pattern and the fact.

Of all the social institutions that have been created, education or schooling is most important because it is the means of perpetuating and improving society. For this reason, it is crucial that students want to go to school. If youth are not interested in what they are being taught, they will not pay attention and they will not learn. Although schools control instruction by selecting what is to be taught and how it is to be taught, students control learning. Ultimately, they control what they will focus on and attend to, and they are prone to attend to those things that interest them. To induce learning, schools must capture and hold the students' attention. Otherwise, young people will not learn to be productive citizens, innovators, or anything else we hope to teach them.

The general approach to interesting students in school is no great mystery. If the instruction students receive in school satisfies their motives, they will want to learn. They will want to attend school and in the process achieve educational objectives. This, of course, is no small order; however, as indicated, such a motive has been identified, the control motive.

The answer to the question, Can the control motive be satisfied in school? is *yes!* Schools are the primary social institutions for developing productive citizens, and as indicated, development of control is essential to productive citizenship. Moreover, the development of any skill requires some exercise of control. Writing a composition requires control of one's handwriting and descriptive ability. In science, control is exercised in conducting laboratory experiments. In math, control is needed to execute the appropriate formula to solve a problem. Additionally, the acquisition and use of knowledge require control; finding facts requires control; and memorizing facts requires control, as does recalling those facts when needed—mnemonic devices are often used to facilitate control of memorizing and recalling facts. So it is both appropriate and necessary for schools to develop students' ability to control.

In contrast, it would not be beneficial to society for schools to provide for the satisfaction of other motives, such as the sex motive. It is patently evident that school is not the place to indulge in sex. Even if it were not illegal, it would be disruptive to learning, and some of the by-products of sex, such as venereal disease and teenage pregnancy, are socially harmful. When sex is dealt with at all in school, students are taught sex hygiene and how to avoid undesirable by-products. The point is that although it is necessary to satisfy students' motives to capture and hold their interest in learning, attempts to satisfy certain motives are undesirable and untenable in school and subvert learning. On the other hand, providing satisfaction for the control motive is compatible with the basic aims of society and schooling. See *Taking Control: Vitalizing Education* (Friedman 1993) for prescriptions for satisfying the control motive in school and teaching students how to control outcomes.

Control is a most prevalent human motivating force that expresses and manifests itself in humans' superior control of their environment. Its importance to human behavior and achievement cannot be overemphasized, although it is often overlooked. In reflection, it becomes clear that people seek to control many of the factors that affect quality of life that were discussed in Chapters 1 and 2, including functional ability and pain. It also becomes evident that all of the improvements discussed in Chapter 3 that contribute to quality of life are attained by finding and developing treatments that will control or bring about the improvements, whether the improvements sought pertain to the preservation or enhancement of their lives. And it is made clear in Chapter 4 that the most advanced level of knowledge and know-how is the control level. The control level subsumes the explanatory, predictive, and descriptive levels. Moreover, the recommended paradigm for achieving improvements is the control cycle (see Figure 4.1).

Most of all, it should become clear that to improve quality of life the control motive must be recognized as the prevalent motive it is, that it can be managed and satisfied without deleterious side effects, and that there is a need to devote more effort to working with it. We now need to consider the human attribute most responsible for superior human control: intelligence.

Chapter 6

Developing Intelligence: The Means to Many Ends

It's quite clear that superior intelligence is largely responsible for improving the quality of human life. Humans' domination in a world of larger, stronger, and more agile creatures can hardly be accounted for without reference to their superior intelligence. And superior mental ability is primarily responsible for the many discoveries and inventions that have enabled humans to prolong their lives substantially, travel to more remote places with increasing speed and safety, communicate greater distances in less time for less money, relieve suffering, repair and replace defunct body parts, master the production of food, clothing, housing, and so on. Although all the people of the world do not as yet benefit from the many remarkable advancements, they are available. And the techniques for reproducing them have been recorded to be shared with the present and coming generations. It's also clear that superior intelligence is in large measure responsible for one person's success in competition with others in a great many walks of life. Intelligent people usually get more of what they want than unintelligent people.

The philosophy of Aristotle keeps cropping up in the literature on quality of life (Nordenfeldt 1993; Megone 1990; Nussbaum 1993). And this doesn't seem to be an accident. He stresses the relationship between intelligence and quality of life in a unique way. He contends that for humans to lead the good life and be happy, they must function in harmony with their essence. The essence of humans, distinguished from other living things, he contends, is their rationality, that is, their ability to reason and to act accordingly (Aristotle 1934). So from Aristotle's perspective, people must act rationally to elevate the quality of their lives. Just instinctively eating, mating, procreating, and foraging is not sufficient for people to fulfill themselves. Accordingly, in Chapter 1, I stressed the importance of rationality, showing that people need not be slaves to their instincts like lower creatures. They can use their superior intelligence to predict the consequences of their actions before they act and act to achieve the goals they want to achieve. No one doubts the superiority of human intelligence. But what is there about human intelligence that is superior to other creatures? The baffling question has been, What attributes of intelligence are responsible for superior human achievements?

Ever since the IQ test was developed by Alfred Binet, IQ has been adopted widely as the primary index of intelligence. Yet it falls far short of predicting superior human achievement. As a matter of fact, it is much more accurate in predicting inferior achievement than in predicting superior achievement. But this might be expected. Binet built the IQ test to identify mentally retarded people so that they might receive the special education they need to succeed in school. It has been a valid instrument for identifying trainable and educable mental retardates ever since. The low end of the scale has the sensitivity to identify people with low functional ability and to make meaningful distinctions among them. Those who score low on the test can be predicted to fail in normal classroom settings and to reach their potentials in special education programs designed for them. But high scores on the test do not predict superior achievement. People who score 160 on the IQ test and are dubbed geniuses have not proven to be more successful than people with IQ scores 40 points lower. Still, there are people who form self-admiration societies because they think otherwise. And high scores on the test are used to select students for "gifted" classes and for other invalid purposes. In the final analysis, only one general conclusion seems to be tenable: People who score very low on the IQ test are mentally retarded and need special education to reach their potentials in school. People who score average or above are not retarded and have the potential to succeed in normal classroom settings.

Another problem with IQ is that it has been conceived of as a stable trait, remaining about the same over time. This suggests that people with low IQs are destined to be mentally inferior throughout their lives and that intelligence is primarily an inherited trait that isn't influenced much by learning. But IQ is a stable trait by ruse of definition, not because intelligence does not increase with learning.

Consider the definition:

$$IQ = \frac{\text{Mental Age}}{\text{Chronological Age}}$$

Mental age increases over time because it is an index of learning. However, it is divided by a higher denominator each year, as people grow older. As a result, IQ remains relatively stable over time. So IQ is stable over time because of the way it is computed, not because intelligence does not increase with learning. Mental age is an index of learning. It increases with learning, and it is just as much an index of intelligence as IQ. It is people's mental age that really indicates their functional abilities, not their IQs. Professionals who use the IQ test to place mentally retarded people in appropriate programs base their placements primarily on the mental ability score of the IQ test because most frequently it is a more accurate index of a person's functional level than his or her IQ score.

Viewing intelligence as a stable trait creates more problems than it solves. It dooms individuals with low IQs, who might otherwise be treated more kindly and helpfully, to be regarded as incurably stupid and groups with lower IQs to be regarded as inferior. In both cases, lower IQ scores might reflect a learning deficit rather than mental inferiority. It has been shown that most people can learn everything that is taught through high school. The difference among people is the amount of evaluation, feedback, and instruction they need over time (Bloom 1968; Bloch and Anderson 1975). Moreover, there is no evidence that people with lower IQs cannot become scientists or chief executive officers (CEOs), given extended instruction. However, there may not be any inclination to give them as much instruction as they may need.

There is a crying need to assess the higher reaches of intelligence responsible for superior human achievement. An attribute of intelligence other than IQ needs to be identified that is possessed by people of superior achievement, such as successful politicians, doctors, military professionals, businessmen, scientists, clergymen, academicians, inventors, and yes, even criminals. An aspect of intelligence needs to be identified that enables success whether a person wants to rob a bank or become a CEO. The attribute need not be a requirement for success at a menial job that does not require exceptional mental ability.

One attribute of intelligence has emerged after prolonged research and development that appears to be responsible for exceptional achievement and improved quality of life—predictive ability. Predictive ability is defined as the ability to accurately *forecast* outcomes from antecedent conditions. It appears to be the mind's major contribution to improved quality of life and other achievements. That is, accurate foresight is the mental factor primarily responsible for success at tasks that require mental competence.

There is compelling evidence that control depends on predictive ability. Industrial psychologists have amassed evidence showing how important predictive ability is to the successful completion of work tasks. Machine operators, as well as other workers, depend on their predictive ability to control their equipment (Leonard 1953; Wagner, Fitts, and Noble 1954; Adams and Xhigriesse 1960). Predictive ability has also been shown to contribute to success in dealing with people (Hayden, Nasby, and Davids 1977). In addition, several studies indicate that good readers are good predictors (Benz and Rosemier, 1966; Greeno and Noreen, 1974; Henderson and Long, 1968; Zinar, 1990), and students with higher predictive ability achieve higher grades in school (Dykes, 1997). Moreover, there is evidence that enhancing predictive ability during instruction increases academic achievement in a number of areas (Walker and Mohr, 1985; Hunt and Joseph, 1990; Reutzel and Fawson, 1990: Chia, 1995; Hurst and Milkent, 1994: Nolan, 1991; Charmello, 1993; Freeman, 1982), and preoperative instructions that enable patients to predict the aftermath of their surgery and how to aid their recovery resulted in earlier discharge from the hospital in comparison to patients who did not receive the preoperative instructions (Healy, 1968). Even play has been cited as a means of experimenting in order to predict and control events (Kelly 1955). For additional evidence, see *The Psychology of Human Control* (Friedman and Lackey 1991).

To appreciate the value of predictive ability in improving the quality of life, it is helpful to understand that it is an attribute that is expressed in many ways. It not only involves the ability to predict that something will happen again because it has a history of happening in a particular pattern; predictive ability also involves being able to predict what might happen in the future that has never happened before and being able to predict ways of making it happen or keeping it from happening. Predictive ability is not exclusively human, but it does develop to a much greater degree in humans. It's all a matter of degree, but it's the degree that matters. Standing erect is partly responsible for the superior foresight of humans; this enables us to see further into the distance. But our big brain contributes much more to our foresight. It enables us to more accurately foretell future occurrences—the supreme benefit of our superior intelligence. The single attribute that distinguishes humans from other creatures and that continues to evolve, providing us with ever-increasing potential for improving the quality of life, is our superior predictive ability.

PERSONAL IMPLICATIONS OF PREDICTIVE ABILITY

The importance to people of predicting the future has been stressed by reminding them that the future is where they will be spending the rest of their lives. Moreover, research has shown that people who score high on a predictive ability test achieved higher job status, were more successful in school, and exhibited greater social competency. In fact, the predictive ability test is a better predictor of success in many instances than IQ tests and other measures of intelligence (Friedman 1974).

Illustrations of the importance of predictive ability to quality of life are easy to find in daily living. Doctors must be able to predict medications that will cure their patients' illnesses; to avoid accidents, drivers must be able to predict how to avoid cars and other obstacles; adults must be able to predict how to invest and save for retirement; parents need to predict child-rearing practices that will benefit their children; and people need to be able to predict how to eat and exercise to remain healthy. Predictive ability enables people to think ahead, plan for the future, and act proactively. They can be much more effective in improving the quality of their lives if they develop and use their predictive abilities to prepare for the future and to shape the future to achieve their ends. If they do nothing more than react to intrusions, they are always at the mercy of their environment, waiting for something to happen to them and recovering from insults and injuries. It is far better to use their predictive ability to be proactive rather than to always be reactive.

The highest form of predictive ability, the form most responsible for improving the quality of life, is *innovative* predictive ability—the ability to predict new ways of enhancing control. Innovative predictions are made by deriving new behavior → outcome relationships. A new behavior → outcome relationship is innovative in the sense that the behavior has never before been executed to achieve the outcome. The execution of the behavior constitutes testing the new prediction that the performance of the behavior will achieve the outcome.

New behavior → outcome relations may be derived in three ways: First, a new behavior may be derived to achieve a known outcome. For example, a new recipe (behavior) is derived for cooking fried chicken (known outcome), a new diet (behavior) is derived for losing weight (known outcome), or a new arrangement (behavior) is derived for playing an old song (known outcome). New behaviors are in essence new procedures. Second, a new outcome is achieved using a known behavior. For instance, legislatures follow established procedures to enact new laws, and the scientific method is followed to make new discoveries. Third, a new outcome is derived and a new behavior is derived to achieve the outcome. The invention of the airplane provides an example. Although individual features of the airplane, such as propellers, wings, and engines, were known before, nothing had been perceived that possessed the combination of features of an airplane (outcome). And the combination of behaviors required to produce an airplane had not been executed previously, although individual behaviors in the combination had been performed before.

Innovative predictions can produce things new to all humankind or just to the person who is innovating. However, it is the production of innovations new to all humankind that is primarily responsible for humans' superior control of the environment.

As exemplified by the invention of the airplane, predicting the achievement of new outcomes and then predicting that newly conceived behaviors will achieve the outcomes seem to be primarily responsible for accomplishments that have markedly improved the quality of life—accomplishments ranging from virtually eradicating polio to the invention of devices such as the X ray, the telephone, and the automobile. This type of sophisticated predictive ability is also responsible for the invention of less celebrated devices that have had less dramatic impact on quality of life such as fasteners. Ingenious innovative predictive ability is responsible for the invention of the paper clip, the hairpin, the button, the zipper, the clasp, the snap, Velcro, the stapler, the safety pin, and other fasteners that people use every day and take for granted.

Humans have the uncanny ability of projecting in their mind's eye future states that have yet to be achieved, planning ways to achieve the states, and following their plans to achieve the states. In being able to predict what may happen, and then predicting ways to make it happen, humans in a very real sense invent the future. They use their predictive ability to invent ways to keep things they don't want to happen from happening. For example, they invent ways to resolve problems that have already happened. They invent medication to cure disease. They even invent ways to improve their inventiveness. They invented the scientific method to enhance discovery.

SOCIETAL IMPLICATIONS OF PREDICTIVE ABILITY

In his seminal work on social structure, R.F. Bales (1950) makes it clear that social structure stabilizes the interaction of group members to reduce uncertainty and unpredictability in the actions of others. So it appears that people form social groups to improve their predictive abilities.

To facilitate predictive ability, societies establish and enforce rules and laws. Enforcement of rules and laws makes behavior in a social group more uniform and thus more predictable. Moreover, the rules and laws enable group members to predict the consequences of their behaviors in social contexts. This enables them to predict that particular behaviors can be employed to achieve the outcomes they want to achieve and the conditions under which others can be predicted to behave in particular ways. Rules of language are defined and enforced to enable members of a society to communicate. Dictionaries specify linguistic definitions, and parents and other teachers teach children to follow the linguistic rules. This enables people to predict the consequences of the statements others may make and to predict the effect that their own statements will have on others. Societal laws specify illegal behavior and so enable people to predict when they and others might be restrained and punished.

In virtually all societies, there are laws and rules against fraud because for a society to exist at all members must do what they promise they are going to do. Otherwise, social relations are unpredictable, and exchanges and cooperation are impossible. For people to exchange and cooperate with one another, lying and cheating cannot be tolerated, even though individuals will try to be deceitful to gain competitive advantage. Fundamental social acts such as appointments require that people meet at an agreed time and place. Marital relations are strained when spouses are unreliable. Unpredictability of people undermines all social relations.

Monetary rules define the value of money, which enables people to predict what they can buy for their money. Societal rules also specify people's social identity by legalizing their names on birth certificates. Thereafter, they learn to predict that they and others will be summoned by their names and expected to respond. In addition, when they sign their names to checks, they are expected to have money in the bank to cover the amounts of the checks. In short, societal rules and laws make life much more predictable and controllable for its members.

As indicated, societies are interested in developing in their citizens sufficient mental competency to take care of themselves and be productive citizens. Mentally incompetent people become social wards rather than contributors. Probably the best index of mental competency is predictive ability. People who cannot predict the effects of their own behaviors on others and the effects of their environments on themselves are indeed mentally incompetent. Furthermore, people who can predict the consequences of their own actions are mentally competent and should be held responsible for their actions legally. For example, people who can predict the effects of drugs on behavior should be legally responsible for committing a crime while under the influence. The implication is that even if they did not know what they were doing while drugged, they are responsible for their actions because they are capable of predicting how they might behave under the influence. Legally defining mental competency in terms of predictive ability clarifies mental competency and guards against definitions of mental competence or sanity used in particular professions, such as psychiatry. Professional jargon may have meaning to members of that profession but may not be translatable to lay language or understood by lawyers, judges, or jury panels.

Predictive ability is also necessary for successful work of any kind, and work is certainly necessary to subsist and improve the quality of one's life. For workers to be proficient at their jobs, whatever their jobs may be, they must be able to predict the effects of their behaviors and the behaviors of their fellow workers on controlling the achievement of the work objective. They may not need to be able to make long-range plans, and they may not be able to anticipate well in other settings—for instance, they may be inept at social gatherings. But to succeed in their specific lines of work, they must be able to make rudimentary, short-term predictions that are accurate.

William Powers (1973a), an engineer, has influenced the behavioral sciences by applying his conception of control theory to explain human behavior. He views control as the means of achieving ends, as many other control theorists do. He does an admirable job describing how mental mechanisms operate to adjust behavior based on feedback to control the achievement of ends. In contrast, prediction theory explains how predictive ability is used to control the achievement of ends and explains further that an end people seek much of the time is the control of outcomes. Predictive ability is the mental ability largely responsible for humans' control of the environment.

The man-machine system highlights the importance of predictive ability to work because it depicts an individual "man" predicting how to control a machine to achieve a specified outcome. Human and systems engineers usually design man-machine systems to achieve a specific objective that contributes to the attainment of overall organizational objectives. For example, a man-machine system was designed to land on the moon as part of our national space exploration program. Given the objectives, systems engineers are able to specify systems functions that must be achieved to fulfill the objective. Then they allocate the functions to the man and machine components.

Likewise, man-man systems depend on predictability. Cooperation requires that workers be able to predict each other's behaviors. To assemble a product, workers must prepare parts of the product according to specifications or the parts won't fit. And the parts must be ready for assembly at an agreed-upon time and place. Production lines require that each worker complete his or her task as specified before forwarding the product. Planning sessions require participants to work together to improve product design. The unpredictability of any team member impairs production.

Commerce of any kind also requires predictive ability. Buyers agree to pay for a product when it has been delivered. Sellers agree to deliver a quality product at a predetermined time and place in order to be paid a specified amount. Contracts are often signed between buyers and sellers to reduce unpredictability, and people are sued for violating contracts. Buyers will not continue to buy from unpredictable sellers. Nor will sellers continue to sell to buyers who don't pay as agreed.

In their book *Competing for the Future*, successful business consultants Gary Hamel at the London School of Business and C.K. Prahalad at the University of Michigan argue that today foresight is the most important ingredient for corporate success. Downsizing, cost cutting, quality control, and consumer awareness are no longer sufficient for success in business and industry. New corporate giants emerge because of their ability to predict how to surpass the competition in the future and how to plan and work in the present to achieve their goals. They offer as an example the Microsoft Corporation's surpassing the International Business Machine Corporation.

In general, predictive ability makes two major contributions to successful production. It enables businesses to predict the products consumers will buy in the future so that they can design and produce products consumers want. Besides, it takes time to bring products to market. Predictive ability also enables businesses to predict how to produce their products as intended. Successful production requires the ability to predict that the execution of particular procedures will produce a product that meets design specifications. This requires establishing quality assurance procedures.

Since predictive ability is related to job success (Friedman 1974), personnel should be screened for predictive ability before employing them to ensure that they have the mental competence to do their jobs. It is also important that personnel development programs increase the predictive ability of personnel. Predictive ability may be upgraded in general or with reference to improving the performance of particular tasks.

Predictive ability is essential to realize the improvements that are sought in any social welfare endeavor. In general, the interest is on predicting treatments that will achieve improvements most effectively and efficiently. Law enforcement officers are interested in predicting ways of reducing crime. Educators are interested in predicting instruction that will improve learning. The medical profession is interested in predicting medications that will improve health. The United Nations wants to be able to predict ways to maintain peace. Welfare agencies want to predict ways of overcoming poverty. And social workers are interested in predicting ways to improve family life.

Statistics are being employed in an increasing number of enterprises to improve predictive ability. Being able to compute probability enables insurance companies to set insurance premium rates that can be predicted to yield a profit. Team managers use statistics to predict player performance in order to increase the probability of winning games. In baseball the earned run averages of pitchers and the batting averages of batters determine when and how players will be used in games. Politicians compile statistics on citizen preferences to predict the slogans that are likely to get them elected. And businesses obtain statistics on customer reactions to their products and services to be able to predict customer satisfaction. Hotels and restaurants routinely ask customers to rate their products and services on questionnaires.

Research and development (R&D) in industry and government involves hypothesis testing to improve predictive ability. The application of scientific methodology requires hypothesis testing. Drug manufacturing companies test hypotheses that new drugs will alleviate symptoms and cure diseases. Automobile manufacturers test hypotheses that new car designs will improve passenger safety. All R&D is aimed at improving the ability to predict the appeal and effectiveness of products and at improving the efficiency of bringing products to market and servicing them.

It behooves societies and organizations within societies to provide for the teaching of predictive ability in schools and on the job. Predictive ability is based on knowledge of antecedent → outcome relations. Knowing such relations enables people to predict that when an antecedent condition occurs, the related outcome will follow. The most useful type of antecedent → outcome relationship that can be taught is the behavior → outcome relationship because it enables people to predict how they can act to improve the quality of their lives. For example, knowing that exercise (behavior) improves health (outcome) enables people to predict that they can improve their health by exercising. Whatever outcomes a society may want their people to achieve to improve the quality of life in the society, they can teach them the behavior necessary to achieve the outcome, enabling them to predict that they can achieve the outcome by executing the behavior. They can also teach people how to be more successful by teaching them how to innovate.

The control cycle can be used to innovate. Within the stages of the control cycle elaborated in Chapter 4, innovative treatments are derived in the stage "Deriving Treatments," after the improvement to be achieved has been projected. Thomas Edison's invention of the incandescent light can be used to illustrate the procedure. The desired improvement might have been defined as an artificial light that burns brighter and longer.

Deriving Treatments

Determining the factors to be controlled. It was known to Edison that particular factors needed to be controlled to produce an electric light: (1) electric current, (2) filament resistance, and (3) burning. Edison could profit from observations of previous inventions, such as the candle and oil, kerosene, and gas lights. These artificial lights produce light from heat. Some only become red hot and produce moderate illumination; others intensify from red hot to white hot and produce brighter illumination.

Discoveries such as Ohm's law preceded Edison and indicated that bright light could be generated from electricity. That is to say, it was known that when increased electrical voltage is induced into a filament that provides substantial resistance, the filament will become white hot and burn brightly.

Determining constraints. Dangerous emissions: Harmful side effects such as dangerous emissions of gas and heat were to be avoided.

Determining means of controlling the factors.

Factor 1: Electric current. Sources of electric current and means of generating increased voltage needed to be provided for.

Factor 2: Filament resistance. Filaments needed to be identified that would resist electric current without disintegrating.

Factor 3: Burning. Means had to be found to prolong the burning of the filaments.

Deriving a treatment. Appropriate sources of electrical current were provided for as well as a means of increasing the voltage of the current. Also, types of filaments were identified for testing that provide resistance to induced electrical current so that the filaments would become white hot and glow brightly. In addition, provisions were made to sustain the illumination for an extended period. This was accomplished by selecting a filament that would not disintegrate as the voltage increased and by encasing the filament in a partially evacuated glass bulb. Removing oxygen from the bulb prolongs the life of the filament. Encasing the filament in a bulb also provides protection from emissions such as gases and intense heat.

Having derived a way to improve control of the factors that needed to be controlled, procedures would be developed to produce an example of the incandescent light as conceived. And a procedure would be planned to assess achievement of the objective. The objective would be achieved if the incandescent light burned brighter and longer without dangerous emissions. The stages of the control cycle would be followed to achieve the objective.

In teaching its members to innovate, societies and organizations are improving their predictive abilities when standard operating procedures do not work, which is primarily in unfamiliar and novel situations. In learning how to innovate, people can predict behaviors to achieve outcomes that have not been achieved previously. This improves their predictive abilities substantially as well as their potentials to improve the quality of life in society.

Another way societies improve the predictive abilities of their citizens is by passing on the lore of the culture from one generation to the next in libraries, data banks, and public records. In propagating the lore, citizens have at hand all the information that benefited the last generation, plus the means of adding to the information to benefit present and future generations.

Still another way societies improve the predictive abilities of their citizens is by teaching them statistics so that they can estimate probability. People learn informally about betting odds, and from casual experience, they develop a sense of when it is likely to rain or snow. Probability is the formal, mathematical way of establishing predictive ability. When people learn how to estimate probability statistically and become aware of its many applications, they acquire a powerful tool for improving their predictive ability and can appreciate how probability estimates are used to increase the success of businesses, industries, sports teams, medical treatments, and the military.

Finally, it can be socially disintegrative to adopt an index of intelligence, such as IQ, that defines intelligence as a stable trait that remains relatively the same over time. This has happened in the United States, where ethnic groups that average lower on the IQ test tend to be regarded as mentally inferior to ethnic groups that average higher on the test. In fact, the ethnic group with the lower IQ may suffer from a learning deficit that can be remedied with instruction. Instruction has been used successfully to increase people's IQ scores. It is also the case that different ethnic groups have different child-rearing practices. Some emphasize education and high test performance. Others regard academics and test performance as peripheral to daily living.

Furthermore, it is difficult to justify the IQ as a selection or placement instrument, since high scores on the test do not predict success in school, on the job, or anywhere else, as mentioned previously. There is no way of justifying the use of the IQ test for rejecting an applicant to any position. And it is unlikely that rejecting an applicant solely on the basis of IQ score would hold up in a court of law. The time-honored way of selecting a person to perform in a position is to identify the demands of the position and test applicants to determine the extent to which they are able to meet the demands. Those able to meet the demands are selected for the position.

Groups cannot justifiably be rejected because on the average they are less able to meet particular demands. In a free society, applicants are supposed to have equal opportunity under the law. Individuals who can meet the demands of a position are qualified for the position, regardless of the groups they may be associated with. Gender, race, ethnicity, color, and creed are not at issue. A woman applying for a longshoreman's job, for instance, should not be rejected because women are not as physically strong as men. If the job demands that applicants be able to lift 200-pound weights over their heads and a particular female applicant can do it, the female meets that job requirement as well as anyone else who can meet that job demand.

For a recent discussion of intelligence, see "Intelligence: Knowns and Unknowns," a report of a task force established by the American Psychological Association (Neisser 1996). The article makes it clear that intelligence test scores alone do not predict success. "Correlations are highest for school achievement, where they account for about a quarter of the variance" (83). Thus, achievement in school is accounted for 25 percent by IQ and 75 percent by factors other than IQ. "Correlations are lower for job performance" (83). So IQ does not account for success in school or on the job. With respect to group differences in IQ, "[t]he range of performance within each group is much larger than the mean difference between groups" (96). The article also makes it clear that schooling can increase intelligence test scores.

The evidence for the effect of schooling on intelligence test scores takes many forms. When children of nearly the same age go through school a year apart (because of birthday-related admission criteria), those who have been in school longer have higher mean scores. Children who attend school intermittently score below those who go regularly, and test performance tends to drop over the summer vacation. A striking demonstration of this effect appeared when the schools in one Virginia county closed for several years in the 1960s to avoid integration, leaving most Black children with no formal education at all. Compared to controls, the intelligence-test scores of these children dropped by about 0.4 standard deviations (6 points) per missed year of school. (87)

BOREDOM AND CONFUSION: IMPEDIMENTS TO PREDICTIVE ABILITY

To be successful in improving the quality of life, people not only need to know how to develop and take advantage of their predictive abilities; they also need to be able to identify, prevent, and recover from impediments that interfere with the improvement of their predictive abilities. To perform well at anything, say, driving a car, people need to develop the required skills as well as to learn how to deal with mishaps. In addition to learning how to start the car, shift gears, and steer it, people need to learn how to deal with breakdowns, accidents, flat tires, traffic jams, and bad drivers. Acquiring these skills enables them to avoid problems most of the time and to solve problems they can't avoid.

Two major impediments to the improvement of predictive ability are boredom and confusion. Confusion occurs when outcomes are too unpredictable—events are too unfamiliar for predictions to be made and tested. And when people cannot comprehend what's happening sufficiently to make and test predictions, improvement of predictive ability is severely impaired. Boredom occurs when events are too predictable—events are too familiar; there is too much sameness over time. The redundancy requires that the same predictions be made, tested, and confirmed over and over again. Under these conditions, it is most difficult to improve predictive ability. Thus, excessive unpredictability causes confusion, and excessive predictability causes boredom. Both are disruptive and disturbing and can severely erode the quality of life.

Boredom occurs when we drive down a straight highway with the same scenery mile after mile, when we watch television reruns we have seen before, when our work becomes too routine, when someone tells us a story we have heard before, when we wear the same clothes too often, when our children keep nagging at us to do something for them, when our diet is too repetitious, and so on. In classic literature, boredom has been illustrated through Nora in Ibsen's *A Doll's House* (1935) and by Walter Mitty in James Thurber's *The Secret Life of Walter Mitty* (1942).

Confusion is also easy to illustrate. We become confused when we begin a new job, travel to a foreign country for the first time, "bite off more than we can chew," acquire a new disease, buy a new house, let work pile up on us, get married or divorced, have a child, spend more than we earn, and so on. Confusion is illustrated in Shakespeare's *A Comedy of Errors* (1936) and in Joseph Heller's *Catch-22* (1961).

Boredom and confusion are so common in our daily lives that there is no need for us to belabor their characteristics. What the reader may not have realized, however, is that most of our psychological problems are either problems of boredom or confusion. We are psychologically disturbed either because events in our lives are too unpredictable and we are confused or because events are too predictable and we are bored. To understand psychological problems more completely, it is necessary to realize that confusion is most often the underlying cause of intense boredom. When people become bored, they have a natural tendency to withdraw from the boring situation to find relief. Boredom becomes more acute when people remain in the situation for an extended period of time. Most often, the reason they remain is because the means of escaping is unpredictable or confusing to them. They feel entrapped in the boring situation because they cannot predict a way out of it without suffering dire consequences. People in boring jobs often find themselves entrapped for financial reasons, and bored homemakers may see themselves entrapped because of their children. In both cases, the people remain in the boring situations because escape is too unpredictable or confusing. So it appears that confusion is most often the underlying reason for prolonged boredom. Therefore, it is the root of all severe psychological disturbances. People with more profound psychological problems tend to be more confused more of the time.

Even though boredom is not deemed to be a serious pathological problem, it can be debilitating. It tends to erode our zest for living and drain our energy; it takes a substantial investment of energy to keep ourselves in boring situations from which we very much want to escape. Moreover, in modern societies, boredom is on the rise. When humans, like lower creatures, spent the major portion of their time battling the fickle elements of nature for their survival, boring occasions were few and far between. However, as survival is ensured, and as more and more leisure time becomes available because of a shorter workweek, increased technology, retirement, and increased longevity, boredom becomes an increasingly prevalent problem.

Still, we must recognize that boredom can have constructive consequences. When we're pursuing novelty to escape boredom, we are most likely to engage in innovative thought that can improve the quality of life. The resulting discovery and invention highlight superior human achievement. Allegedly, Archimedes was idling in a bath, Newton in an orchard, and Einstein in a patent office at the time they discovered buoyancy, gravitation, and relativity, respectively.

Boredom and confusion can be situation specific; and it is possible to be bored in one situation and confused in another. We might be confused at home because of our inability to predict the behavior of our spouses or children and bored at work because of the monotonous routine imposed by our jobs. Moreover, people react differently to boring and confusing situations from time to time. When people are energetic and alert, they frequently are interested in improving their control. Under these conditions, boring or confusing stimulation will irritate them. On the other hand, when they are fatigued and interested in resting, boring stimulation, such as counting sheep, will calm them and put them to sleep. Confusing stimulation would tend to be irritating because it might prevent them from resting.

Personal Implications

One of the advantages of a theory is that it explains why we have a problem and helps us to solve it. Prediction theory explains the causes of boredom and confusion and implicitly suggests remedies for them. Since boredom is caused by too much predictability, relief can be achieved by introducing novelty and challenge into our lives. So even if we find ourselves stuck in a boring situation, we can get relief by engaging in novel activities and meeting new challenges to relieve the boredom and enable us to eventually extricate ourselves from the situation. For example, while enduring boring jobs, people can learn new trades that interest them and eventually become employed in them. There is considerable evidence indicating that bored people seek novelty (see Smith, Meyers, and Johnson 1968; May and Hutt 1974; Brickman and D'Amato 1975) Since confusion is caused by too little predictability, relief can be achieved by introducing familiarity and stability into our lives so that we may be able to predict outcomes again. For instance, people can reduce novelty and retreat to the familiar. There is substantial evidence that confused people seek familiar and redundant stimuli (see Bindra 1959).

The situations people seem to prefer are optimally predictable situations. When situations are optimally predictable, they are familiar and stable enough to predict outcomes and novel enough to allow us to improve prediction. When outcomes are either too predictable or too unpredictable, people are disturbed. This can be conceptualized as a bipolar continuum with optimum predictability in the center and boredom and confusion at either end (see Figure 6.1).

When we find outcomes to be too unpredictable and we are becoming confused, we must seek familiar, stable conditions to find relief. When we find outcomes to be too predictable, we must seek novelty and challenge. The contention that too much and too little predictive ability are disruptive is supported by the research of London, Schubert, and Washburn (1972). Although optimum predictability may only be personally recognizable as the absence of boredom

Figure 6.1
Bipolar Continuum

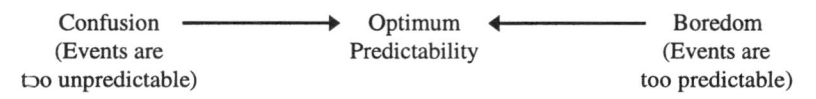

Confusion ——————▶ Optimum ◀—————— Boredom
(Events are   Predictability   (Events are
too unpredictable)   too predictable)

Source: Author.

and confusion, it and the bipolar continuum have theoretical importance in enhancing conceptualization of boredom and confusion and the direction we must move to find relief from them. The bipolar continuum helps us see that when we are moving toward optimum predictability to find relief from confusion, we can go too far and induce boredom by removing too much novelty from our lives. Conversely, when we are moving toward optimum predictability to overcome boredom, we can go too far and induce confusion by introducing too much novelty. In a way, optimum predictability represents an elusive "golden mean" we should attempt to approximate. Support for this interpretation of boredom and confusion and the bipolar continuum is provided in the book *Human Nature and Predictability* (Friedman and Willis 1981).

Societal Implications

Societies can improve the quality of life of their citizens by teaching them how to prevent, diagnose, and find relief from boredom and confusion as follows.

Preventing Boredom and Confusion. To prevent confusion, people must plan to avoid being in situations that are too unfamiliar for prolonged periods of time. Before people become involved in unfamiliar situations, they need to learn about them. To prevent boredom, they must plan to avoid being in monotonous situations for prolonged periods of time.

When people make plans, they need to recognize that boredom and confusion can undermine their efforts. They need to make and review their plans before taking action to make certain that they are not going to get bogged down in boring or confusing situations. As they make continued attempts to improve the quality of their lives, they need to steer a course between boredom and confusion, an optimally predictable course, that is, a course that provides sufficient novelty for them to avert boredom but not so much novelty that they become confused.

Diagnosing Boredom and Confusion. As hard as they try, people will not always be able to avoid boredom and confusion. Consequently, they must learn how to recognize and diagnose boredom and confusion so that they can deal with them effectively. The first thing they will recognize when they are becoming either bored or confused is that they are disturbed, and they may not know immediately whether it is because of boredom or confusion. To determine whether they are disturbed because of boredom or confusion, they need to ask themselves the following questions.

Are the same things happening over and over in the situation?
Is there too much sameness?

If the answer is yes, they are bored with the situation.

Is there more going on than they can handle effectively in the situation?
Is it difficult for them to know what to expect?

If the answer is yes, they are confused with the situation.

Finding Relief from Boredom and Confusion. Since boredom occurs because events are too predictable, boredom can be relieved by introducing novelty and challenge in people's lives. Since confusion occurs because events are too unpredictable, confusion can be relieved by simplifying and regulating their lives and by engaging in familiar activities.

Sometimes they will need to tolerate boredom or confusion for limited periods of time. They may need to tolerate a boring teacher. Or they may need to tolerate the confusion that occurs when they are trying something new, say, a new game, until they become more familiar with it. At other times, they may be able to relieve their disturbance by applying immediate solutions, such as doing something novel if they are bored or retreating to a familiar, stable environment if they are confused. If their disturbance is intense and they do not find an immediate solution, they will need to systematically attack the problem. They may need to seek help or make long-range plans. If they are bored, the objective they pursue is moving from a state of boredom to a state of optimum predictability. If they are confused, the objective they pursue is moving from a state of confusion to one of optimum predictability. When they make plans, they need to keep in mind that if they introduce too much novelty to relieve boredom, they will create confusion; or if they remove too much novelty to relieve confusion, they will create boredom. They must always try to approximate optimum predictability when they are striving for a solution.

In treating maladies of the body, cause-effect relationships between treatments and ailments are often quite clear. For example, a bacteria or virus is known to cause the ailment that is diagnosed, and the treatment is known to cure the ailment by killing the bacteria or virus. Such is the case in using Biatrin to kill *H. pylori* bacteria to cure ulcers. Similarly, prediction theory reveals the cause-effect relationships between basic psychological ailments and their treatment. In a nutshell, mental stress is caused by boredom, too much predictability or confusion, too little predictability. Boredom is treated by introducing novelty and challenge to reduce predictability. Confusion is treated by involving people in more stable, familiar endeavors to increase predictability. Although the diagnosis and treatment of mental ailments are much more complex than this, the fundamental cause-effect relationships underlying the diagnosis and treatment of psychological ailments are clearly explained by prediction theory. More complex diagnostic-treatment relationships might well be better understood within the context of prediction theory.

Presently, the cause-effect relationship between the diagnosis and treatment of far too many psychological ailments is either nonexistent or unclear. Diagnostic designations such as "attention deficit disorder" and "depression" are amorphous, unclear, and not helpful in prescribing effective treatment. Attention deficit disorder (ADD) as a diagnostic designation includes all people who have difficulty paying attention, a vast multitude of people. The diagnosis indicates neither the cause, which might be almost anything, nor the appropriate psychological treatment. In fact, the treatment is quite often medication rather than a psychological treatment. The common treatment for ADD in children is Ritalin. And parents and teachers who are having difficulty managing children are prone to have them diagnosed as ADD and to have Ritalin prescribed for them to make them more manageable. There appears to be less knowledge or concern about the side effects of the drug on children, especially its interference with learning, the development of autonomy, and appropriate psychological treatment. Attention deficit disorder might be caused by either boredom or confusion. The psychological treatments for boredom and confusion are quite different, as has been explained.

Depression is another designation of a psychological ailment that is unclear and for the most part useless. A television commercial beseeches people who are suffering from depression to get help from their clinic or somewhere else. Numerous symptoms of depression are listed that are so common that many, if not most, of the viewers might identify with one or more of them and become worried. Generally speaking, depression as a diagnostic designation includes all people who are dejected at the time for one reason or another, whether the dejection is transient or inconsequential to their daily living. Although some clinicians say that they can distinguish between nonvirulent and virulent or clinical depression, no clear distinction between the two has been established. Moreover, the diagnosis does not indicate the cause—again, it might be almost anything—nor the appropriate treatment. Treatment may be protracted and elaborate if the person can afford it or has sufficient insurance coverage. Or it may be as simple as a drug prescription such as Prozac. Here again, the drug treatment might interfere with mental acuity and appropriate psychological treatment. Depression caused by confusion requires one kind of psychological treatment, while depression caused by intense, inescapable boredom requires another kind of psychological treatment. Using drugs to relieve psychological symptoms does not preclude the need for psychological treatment to cure the psychological ailment and to foster autonomy.

To achieve societal goals, it is necessary to keep citizens from becoming too bored or too confused. Schools need to provide optimally predictable learning environments. Otherwise, students will not be able to achieve educational objectives. Optimally predictable learning environments provide conditions that are sufficiently familiar to students to enable them to make predictions and sufficiently novel to challenge them. Since new learning is built on prior learning, it is essential that students be familiar with materials upon which new learning is based, or they will be unable to learn. For instance, if they are not familiar with addition, they will not have the foundation for predicting solutions to multiplication problems. So the first condition for an optimally predictable learning environment is that learning conditions not be too unpredictable.

The other condition for providing an optimally predictable learning environment is that learning conditions not be too predictable. The learning environment must contain sufficient novelty to challenge students and enable them to improve their abilities to predict. Students must be confronted with new material to challenge them and maintain their interest in learning. Too much predictability in learning is often caused when students are required to continue to study material they have already mastered. Such might be the case if after students mastered multiplication, they were given an additional 20 multiplication problems for homework.

Traditional bureaucracies undermine the effectiveness of personnel. They create confusion by requiring people to traverse a maze of red tape to get anything done. In addition, rules may prevent people from succeeding. People are required to go through formal channels. Yet they know that the best way to succeed is to obtain informal access to their superiors. People are told to expose obstacles to the success of the organization. But if they try to blow the whistle on an inept superior, they jeopardize their jobs. There are too many changes too often. Every time executives become aware of a vulnerability, they issue a memorandum to cover themselves. People drift into the habit of protecting themselves as a first priority at the expense of achieving organizational goals. People often cannot get their job assignments done because they are not given the authority necessary to carry out their responsibilities. There are also inherent contradictions in expectations of people. They are expected to cooperate with people they are in competition with. And there are often contradictions in sacred and secular values imposed on them. During the week in the free enterprise system, it's dog eat dog. On the weekend in houses of worship, people are told to love and to be their brother's keeper.

Bureaucracies also generate boredom, however inadvertently it may happen. People are required to make the same report on a number of different forms and to different people. They are sold the "company line" over and over and are expected to espouse it openly to show that they are team players. They are often assigned repetitive tasks to perform over prolonged periods without sufficient diversion. Keyboard operators, filing clerks, and production line workers frequently suffer from boredom. Bosses often repeat themselves to make their point, to assert their authority, and to keep on asserting their authority. Many work environments produce the same sights and sounds monotonously. Machines tend to go through the same motions and produce the same sounds repetitiously. There are numerous other examples that could be mentioned. Productivity and the quality of people's lives in the workplace can be considerably improved if sources of boredom and confusion are reduced.

In summary, the mind's primary contribution to improving the quality of life is predictive ability, especially innovative predictive ability. Societies can contribute to the improvement of the quality of life of their citizens if citizens' predictive abilities are developed. As a result, they will be better able to fulfill their own aspirations and make social contributions. Individuals will get more of what they want from their lives if they develop and use their predictive abilities to predict behaviors to improve the quality of their lives, test their predictions, and revise their predictions based on feedback. This thoughtful, proactive behavior will do much more toward improving the quality of people's lives than reacting to environmental intrusions or following instinctive prompts impulsively. Finally, the quality of life can be improved substantially if boredom and confusion are reduced. To begin with, predictive ability needs to be recognized as the aspect of intelligence largely responsible for improving the quality of life in the past and a most valuable resource for improving the quality of life in the future.

Predictive ability is essential to improving the quality of life in all eight areas described, both on a personal level and on a societal level.

On a personal level. In the area of government, people need to be able to predict how to avail themselves of their legal rights and entitlements. In the area of health, people need to be able to predict how to improve their strength, endurance, mobility, interpersonal relations, and mental stability, as well as what they must do to recover from mental and physical ailments. In the area of work, people need to be able to predict how to increase their earnings and work satisfaction. In the area of education, people need to be able to predict how to increase their knowledge and skills. In the area of remote access, people need to be able to predict how to travel and to transport merchandise to and from distant places. In the area of protection, people need to be able to predict how to ensure their own and their loved ones' security. In the area of provisions people need to be able to predict how to acquire the products, services, and housing they need.

On the societal level. In the area of government, societies need to be able to predict that their government will help ensure their survival and optimize their free choices and political opportunities. In the area of health, societies need to be able to predict how to increase the longevity, functional ability, and contentment of their members on the average and decrease the rate of occurrence of controllable diseases. In the area of work, societies need to be able to predict how to increase the purchasing power of their members. In the area of education, societies need to be able to predict how to increase the percentage of their members that are educated and the average level of their education. In the area of remote access, societies need to be able to predict how to increase the average number of messages and shipments sent to and received from distant places and the average distance traveled by their members. In the area of recreation, societies need to be able to predict how to increase the rate of recreational involvement of their members. In the area of protection, societies need to be able to predict how to decrease the rate of crime and foreign incursion. In the area of provisions, societies need to be able to predict how to increase the number and variety of products, services, and types of housing available to their members. In short, it is the predictive ability of people individually and collectively that is largely responsible for humans' solving quality of life problems.

Most of all, it is becoming eminently clear that we need to change the way we think about intelligence. Predictive ability needs to be recognized for what it is, an attribute of intelligence largely responsible for superior human achievements, an attribute that can and should be developed through instruction. If we are to improve the quality of people's lives, we need to concentrate on designing instruction to develop individuals' predictive abilities to their highest potentials. The fact that there may be individual and/or group differences in predictive ability and other attributes of intelligence is beside the point and a far less important field of inquiry.

Chapter 7

The Holistic Approach

It was said early in the book that a holistic approach to quality of life issues has great advantages. From a holistic perspective, improving the quality of life is the overall or whole problem being addressed in this book, and factors that effect improvement are parts of the problem. To be effective in improving quality of life, it is necessary to understand relevant parts/whole relationships. Professionals that focus on component factors in isolation lose perspective and context and therefore must suffer in their efforts to understand and help people. And professionals who try to deal with overall quality of life without considering the factors that contribute to it are being superficial and also must suffer in their efforts to understand and help people.

The holistic approach gains justification from Gelstalt psychologists who have proven quite conclusively through their research the advantages of understanding the relationships between parts of a problem and the whole in solving the problem (Kohler 1925, 1929, 1969). It will be enlightening, useful, and befitting at this time to knit the parts of the problem and solutions discussed in previous chapters into a meaningful whole. This synthesis can also serve as a summary of important factors discussed in the book.

The major divisions of factors to be dealt with in improving quality of life are structural factors and dynamic factors. Structural factors refer to the static factors; dynamic factors refer to interactive factors. All other factors discussed will be parts of the two major divisions. We will first consider structural factors.

STRUCTURAL FACTORS

The presentation of structural factors enables one to see the important quality of life issues in relationship to one another. To improve quality of life, these factors must be understood and dealt with. Three major factors that have been discussed previously and need to be considered when attempting to improve quality of life are (1) the improvement sought, (2) the area of concern, and (3) the level of concern.

The Improvement Sought

To help people, it is important for professionals to understand the nature of the improvement they are trying to achieve. Understanding the relationship between two types of improvement, preservation and enhancement, can be very helpful.

Preservation. Preservation was discussed extensively in Chapter 1. It pertains primarily to sustaining life. Whenever people's ability to maintain control of outcomes sufficiently to survive and cope with the contingencies of life is compromised, an improvement sought is the preservation of control. Preservation of control is concerned with many issues and is attended to in many different ways.

Malnutrition is a major threat to preservation of control. As we know, without replenishing depleted nutrients, living things wither and die. Humans not only forage to refuel like other creatures; they continually attempt to improve the quality of their diets. Two diet pyramids have been proposed to help people maintain a healthy diet. In 1992, the U.S. Department of Agriculture (USDA) published the "Food Guide Pyramid" presenting the government's conception of a nutritious diet. The emphasis is on the consumption of grains, fruits, and vegetables, with a moderate amount of dairy foods and little in the way of fats, oils, and sweets.

In the summer of 1994, the World Health Organization, the Harvard School of Public Health, and Oldways Preservation and Exchange Trust, a dietary think tank, recommended a different dietary pyramid, the "Mediterranean Diet Pyramid." In contrast, the Mediterranean pyramid differentiates between less desirable saturated fats and unsaturated fats as well as between red meat and other meats. It also recommends more beans and nuts and less meat. It favors fish and poultry over red meat and eliminates butter and margarine. Moreover, it allows people who are not overweight to add some olive oil to their diets. The Mediterranean diet is an attempt to refine and improve on the USDA's diet, profiting from knowledge of the diets of people living on the island of Crete and other Mediterranean cultures who live longer and who are less likely to die of coronary disease than Americans.

Many dietary experts claim that people can get sufficient nutrition without dietary supplements, but few see any harm in taking a multivitamin/mineral supplement. Furthermore, an increasing number of experts are recommending calcium supplements, realizing how difficult it is to get sufficient calcium from a normal diet and seeing the long-term risks of calcium deficiency.

Atrophy is another threat to the preservation of control. A particular adage keeps cropping up that is becoming more and more credible as research findings accumulate: "Use it or lose it." It is becoming increasingly clear that an inactive, sedentary lifestyle is dangerous. In addition, if mind and/or body functions are to maintain optimum proficiency, they must be used, even taxed, regularly. The evidence that people must use or lose body functions is substantial. Evidence that people must use or lose mental functions is mounting but not yet conclusive.

Of course, exercise is the prescription for preventing and overcoming atrophy. Prevention of physical atrophy can be achieved by performing physical exercise. In general, physical atrophy can be prevented by building up gradually to 50-minute periods three times a week. A single exercise is sufficient if it exercises both the upper and lower body and provides aerobic benefits all at the same time, for instance, the use of a cross-country skiing machine.

A holistic prescription for maintaining physical fitness can be proffered combining prescriptions for diet and exercise, adding a prescription for weight control as a guiding precaution, because it is quite possible to follow healthy diet and exercise regimens and become unhealthy because of overweight or underweight. It should be mentioned that a recent study indicates that it is healthier to use slimmer weight standards (lower weight for a given height) (Manson et al. 1995). For additional information on factors that affect physical fitness and health, see the position statement on exercise of the American Heart Association (1992), the year 2000 health objectives (*Healthy People 2000* 1991), and Paffenberger et al. (1986).

People who maintain their physical fitness over the years retain a great deal of their functional abilities. Many still accomplish amazing feats well into their declining years. People over 70 still run in 26-mile races, and a 70-year-old marathon champion can run the marathon 70 percent as fast as a world-class young runner (Moore 1975). At age 72, Christo Varzakis completed a marathon in under four hours. At 79, after running in competition for more than 63 years, he showed no signs of the chronic disease typically associated with aging (Rontoyannis 1992). Amazingly high levels of strength are evident at very old ages. A 74-year-old Bulgarian weight lifter bench-pressed 341 pounds (Spirduso 1986). And many well-trained men at 60 have the same functional capacities as 40-year-old men (Skinner, Tipton, and Vailas 1982).

There is so much written on diet, exercise, and weight control that it is difficult to see the forest for the trees. However, guidelines for maintaining physical fitness are frequently made more detailed and complicated than need be. Given the present knowledge base, people can maintain physical fitness by following the Mediterranean diet, exercising on a cross-country skiing machine for approximately 50 minutes three times a week without overexertion, and keeping their weight within preferred risk bounds. Why isn't the maintenance of physical fitness conveyed to the public in simple, holistic terms?

Recovery from physical atrophy can also be achieved through exercise, even for people who have led a sedentary life and have been inactive for a long time. After 36 weeks in an exercise program, sedentary men gained as much strength, endurance, and mobility as men who had previously been college athletes (Brill et al. 1989). Further, people can reap great benefits from a fitness program even if they are over 60. Sixty-year-olds can increase the amount of work they can do (as indicated by maximal oxygen consumption) through training almost as much as 30-year-olds (Hodgson and Buskirk 1977).

Even if they have lost vital functions late in life, people may be able to make a comeback. Research was conducted on people from 86 to 96 in aging centers and hospitals in Boston who could not walk without a cane or rise from a chair without using their arms. After two months of lifting and lowering their legs with weight resistance, 24 times, 3 times a week, their leg strength increased 3 to 4 times. And their walking speed increased 48 percent ("Pumping Iron" 1990). Other losses due to aging can be regained as well. One perplexing problem of aging that impairs performance in so many activities from sports to driving a car is that coordination and reaction time slow down. However, this trend can be reversed. The coordination of older individuals who were aerobically trained for four months was significantly faster (Dustman et al. 1984).

These are just a few of many research studies that show, that physical conditioning can help prevent premature debilitation and enable elderly people with diminished capacity to regain strength, stamina, and agility. For further corroboration, see Kasch et al. (1990), O'Brien and Vertinsky (1991), Chick et al. (1991), Moritani and deVries (1980), Ostrow (1984), Hart (1981), Sherwood and Selder (1979), Spirduso and Clifford (1978), and Dustman et al. (1984).

Ailments are a third threat to preservation of control, primarily illness and injury. Physical ailments can be prevented to some extent by following a physical fitness regimen. Physically fit people are, generally speaking, less prone to illness and injury. Other safeguards, of course, are being vaccinated against polio, smallpox, and other dread diseases; saying no to harmful drugs; and taking precautions to avoid injury. Taking precautions to avoid injury ranges from wearing a seatbelt while driving to avoiding prolonged exposure to the sun to avoiding dangerous neighborhoods.

Prevention of psychological loss of control can be ensured by preventing extensive boredom and confusion from encroaching on one's affairs and developing the predictive ability necessary to cope with unfamiliar situations before becoming involved in them. Boredom can be prevented by avoiding prolonged involvement in situations that are too familiar and redundant. Confusion can be prevented by avoiding extended exposure to situations that are excessively variable and unfamiliar. Developing predictive ability to deal with new situations involves learning through instruction. Confusion can be prevented by learning about a foreign country before going there for the first time.

Recovery from ailments is another matter. Some ailments have proven to be quite difficult to recover from, drug addiction in particular. People might have a genetic predisposition to become addicted. And once getting hooked on addictive drugs, both physical craving and psychological dependence need to be overcome. Some illnesses people can't recover from. Some diseases are fatal;

others people must learn to live with. AIDS is often fatal, and people learn to live with loss of control of body functions as they grow older. People can undertake the prevention of illness and disease on their own. Recovery quite often requires the intervention of a trained professional, such as a doctor.

Recovery from psychological ailments requires some form of psychotherapy. From the vantage point of prediction theory, severe thwarting of the control motive causes despondency and depression. Loss of control of intake, mobility, productivity, and decision making can be devastating, as explained in Chapter 5. And there are traumatizing changes in life when control can be threatened, such as marriage, divorce, being hospitalized, and becoming a parent for the first time.

Psychological trauma is reduced when people become able to predict that they will be able to improve their control. Relief does not necessarily require improvement of control but rather the ability to predict that control will be improved or at least that control can be improved. It is predictive ability that is crucial in reducing psychological trauma. The ability to predict that improvement may be forthcoming is the essence of hope.

In Chapter 6, we discussed why predictive ability is so essential to personal security and mental health and why boredom and confusion can generate psychological trauma. The means of recovering from boredom and confusion were also explained. As indicated, relief from confusion is gained by finding a more predictable environment; relief from boredom is gained by finding a less predictable environment.

There appears to be a natural psychological defense mechanism that engages when confusion or boredom become too extreme, causing pronounced psychological trauma. In the case of extreme confusion the unpredictable stimulation causing the trauma is mentally shut or gated out of awareness. In the case of extreme boredom the monotonous stimulation causing the trauma is mentally shut out of awareness. Defensiveness is adaptive in that it enables people to recover from the debilitating effects of psychological trauma. It is maladaptive because the traumatizing event being shut out of awareness cannot be dealt with. (See Friedman and Lackey [1991] for more on defensiveness.)

Once sufficient complacency has been restored so that the traumatic event can be dealt with, people become preoccupied with it in an effort to be able to predict it. If they can predict the event, they can avoid it, even if they can't control it as they would prefer. This preoccupation manifests itself in what Freud called the "repetition compulsion," which is the need to consider a traumatizing event repeatedly until it can be predicted.

Jennifer Savitz (1979) has derived a successful psychotherapy from prediction theory. In it she diagnoses boredom or confusion, as the case may be, and provides for immediate relief by engaging bored people in novel activities and confused people in simpler, more familiar activities. She then teaches them how to use the control cycle to make and achieve long-range plans. This teaches them to solve their own problems, makes them more independent, and helps wean them from psychotherapy.

Treatments to alleviate psychological ailments are many and varied, and they are used to cure physical ailments as well as mental ailments. Support groups are fashionable and effective when they include people with common ailments such as cancer. Knowledge gained from participation in support groups can increase the ability to predict consequences of the common ailment, as well as aid in controlling consequences. Other fashionable treatments include relaxation therapy, biofeedback, hypnosis, meditation, conversational therapy, drug therapy, and electric shock therapy. Each claims to relieve ailments of one kind or another. Many do not provide for development of autonomy, as Jennifer Savitz's method does. They do not develop in the patients the ability to predict and control their own lives.

Fatigue is a fourth threat to the preservation of control. Attempts to exercise control are fatiguing. Periodically, people must take time to recuperate from fatigue so that they can resume pursuit of control with renewed commitment, acuity, and vigor. Rest and recreation are among the most common forms of recuperation. Rest may involve a full night's sleep, a nap, quiet repose in an easy chair, soaking in a warm tub, or a fifteen-minute break. Recreation may be passive or active. Passive forms of recreation include watching television, chatting on the phone, doing crossword puzzles, collecting stamps, and attending sports contests. Active forms of recreation range from mildly strenuous activities such as golf, Ping-Pong, and bowling to more strenuous activities such as weight lifting, karate, and boxing. Quite often, recreation is used to overcome the fatigue caused by boredom or confusion resulting from work.

Compensation is often needed to preserve control. When people can't recover from losses or disabilities that impair their control, they need to try to compensate. As technology becomes more successful in saving and prolonging the lives of disabled people who have lost functional capabilities, it becomes increasingly important to compensate for the losses to restore functions. This is an area that deserves and probably will be given more attention in the future. Prosthetics is a burgeoning field providing more sophisticated artificial replacements for body parts ranging from tooth replacement to the replacement of joints such as hips and knees. As prosthetic replacements improve, so do the essential body functions they abet like eating and walking. Other replacements cater more to human vanity than necessity. Hair replacement continues to be a big industry offering an increasing number of options.

Compensatory mechanisms keep improving, too. Ambulation and mobility improve with the refinement of both hand-driven and motor-driven wheelchairs. Motor scooters also are becoming more varied and specialized for different purposes. And cars and vans can be adapted to compensate for a greater variety of disabilities. And, of course, there are other aids such as canes, crutches, and walkers that facilitate ambulation, glasses to compensate for failing vision, and hearing aids for failing hearing.

The Holistic Approach 163

Moreover, the disabled are taught how to maintain and regain vital functions by developing and adapting an able body part to compensate for a disabled body part. People who have lost their legs or the use of their legs are routinely taught to use their arms to partially compensate. And people who have lost their arms or use of their arms are taught to use their legs and toes to partially compensate.

Preventive compensation involves more than being taught how to use body parts differently. It involves such compensatory measures as learning to increase illumination to compensate for failing eyesight and installing hand rails to steady movement. Skin moisturizers can be used to compensate for drying skin. An electric bed can be used to help people get out of bed, and an electric chair can assist them in getting out of a chair. Support hose can be prescribed to aid circulation and steady walking. Food can be supplemented to prevent and compensate for ailments of aging. For example, hormone replacement can be used to compensate for hormone depletion, to name a few compensatory measures that can help.

People will quite naturally attempt to compensate for impaired functions. Unstable walkers will walk more slowly and take smaller steps. Pregnant women will walk leaning backward to compensate for the increasing anterior weight of the growing fetus. And people who are losing their hearing acuity will rely more on lip reading to understand spoken communication. To understand the subtleties and ingenuity of compensation, it is only necessary to have elderly people who live alone show one all of the arrangements they have made in their surroundings to compensate for their impairments. However, the challenge for technology is to improve on compensatory aids.

Mental failures can be compensated for as well as physical failures. Loss of memory need not be as devastating as it often is. Memory joggers can be used to compensate. Elaboration is the key to committing things to memory, as well as associating the unknown to something already known. For example, to remember a number, say, 1,812, an individual can associate it with something he or she knows such as the War of 1812. One can elaborate it by writing it down and repeating it. A test of failing memory and mnemonic techniques to compensate for forgetfulness have been published in Consumer Reports on Health (1995).

This concludes our discussion of improvements that can be made to preserve life. For additional views on the promotion of health, see Downie, Fyfe, and Tannahill (1990). We will now consider improvements that enhance life. These are related but different kinds of improvements.

Enhancement. As indicated in Chapter 1, enhancement extends beyond preservation. Although preservation is a potent innate drive, and people will attempt to preserve what they have as best they can, they will also attempt to enhance their lives. They want their lives to be better in the future than they are in the present. And their aspirations take many forms and are expressed in many ways. Aspirations may be as simple and concrete as increasing functional ability above advocated acceptable standards for Americans. (In Chapter 2, I commented that there are advantages in achieving superior functional ability. And I criticized existing functional ability scales for measuring only disability or its absence, an indication that superior functional ability is being given short shrift.) Other concrete enhancements people might aspire to are a larger home, a better car, finer clothes, and more money. People also seek more abstract forms of enhancement. They may aspire to reaching greater peace of mind, attaining nirvana, going to heaven, or achieving happiness.

Our concern is not so much with how people hope to enhance their lives but with the fact that everyone does in one way or another, almost all the time. A better understanding of enhancement can be attained analyzing the pursuit of happiness as an example of an enhancement people pursue.

The pursuit of happiness is an unalienable right Americans are endowed with by the United States Constitution. Happiness, however, is conceived of differently by different eminent thinkers. Thomas Jefferson offers guidelines. He suggests that happiness is to be not pained in body nor troubled in mind. Views of happiness attributed to others include: health, peace, and competence (Alexander Pope); good friends, good books, and a sleepy conscience (Mark Twain); little advantages that occur everyday (Benjamin Franklin); to find out what one is fitted to do and to secure the opportunity to do it (John Dewey); serving others (Albert Schweitzer); a quiet life, for it is only in an atmosphere of quiet that true joy may live (Bertrand Russell).

In reading a list of attributes of happiness, it is easy to see that happiness means different things to different people. And we have many ways of alluding to this inconsistency: One man's meat is another man's poison; different strokes for different folks; or as Amy Lowell said, "Happiness, to some elation; is to others mere stagnation." But this great diversity of meaning among people need not be a problem. It might merely be a recognition of the vast individual differences among people. Still, the fact that different things make different people happy and that it takes different things to make the same individual happy at different times need not interfere with our quest for an understanding of enhancement, provided enhancement has some common attribute for all people.

But are there attributes of enhancement common among people? Is there similarity as well as diversity? Prediction theory provides an explanation worth considering. The theory acknowledges that different people want different things and that individuals want different things at different times. But it contends that all people want to improve their control of themselves and their environments. Although they may have achieved as much control as they care to achieve in particular areas, they are always interested in improving their scope, degree, or reliability of control in some area. They don't want to stagnate. Furthermore, whatever people want at any given time, they prefer to control its acquisition as opposed to just having it. This enables them to have it whenever they want it. People will attempt to accommodate to limitations and threats when they need to, but they prefer to control the sources of their satisfaction and will attempt to control them when they have the opportunity.

Translated in terms of happiness, we might say that whatever it takes to make people happy, they want to control it. This makes happiness a consequence of gaining and exercising control. Thus, control is essential to happiness for all people, whatever individuals may need to control to make them happy. Others have acknowledged that happiness is a consequence or a by-product. As Aldous Huxley is alleged to have said, "Happiness is like coke—something you get as a by-product [of coal] in the process of making something else."

Prediction theory adds a very essential dimension to our understanding of happiness as a by-product. It explains that happiness is a by-product of control. Whatever else people may need to be happy, ultimately happiness or any other type of enhancement will be a consequence of their being able to predict the control of its acquisition.

This concludes the discussion of factors that need to be considered in determining the improvement to be sought, both preservation and enhancement factors. It should be noted that preservation most often takes precedence over enhancement, because it is necessary to sustain life before it is possible to enhance life. Moreover, prediction and control are required to preserve life or enhance life effectively.

Finally, pursuit of improvement is proactive; people work in the present to improve their lives in the future, whether they are working for preservation or enhancement. Psychologically speaking, this is a healthy orientation. In contrast, preoccupation in the present with past events is psychologically unhealthy, and in the extreme, it can be pathological. Being fixated in the present with past events can preclude the pursuit of improvement. The elderly quite naturally reminisce about the good old days when most of their lives are behind them and they have little to look forward to in the future. However, they are better off working in the present to improve their control in the future as much as they are able to. Severely guilt-ridden people are preoccupied in the present with sinful acts they believe they have committed in the past and can be bogged down in the present by the weight of their guilt feelings and efforts to undo their sins. They often need psychotherapy to help them overcome their fixation.

Area of Concern

The areas that must be considered in any effort to improve the quality of life have been thoroughly discussed in Chapter 3. They are government, health, work, education, remote access, recreation, protection, and provisions. An improvement sought most probably will be concerned with one area more than another. However, to maximize achievement, it is important to determine whether factors in any or all of the eight areas can contribute to the achievement of the improvement. The selection or design of a treatment to achieve a desired improvement may be limited and simple, or it may be multifaceted and complex.

Table 7.1
Quality of Life Database Format

	Societal		Personal	
Areas	Improvements	Treatments	Improvements	Treatments
Government				
Health				
Work				
Education				
Remote Access				
Recreation				
Protection				
Provisions				

Source: Author.

Although all improvements sought need to be defined and may pertain to one particular area, it may well be that the prescription of an effective treatment will require intervention in a number of areas. Consider, for example, that the improvement sought is the reduction of drug-related mortality and morbidity. The improvement sought may center more in the area of health than any other area. However, the treatment would probably involve a number of areas. Education might be used to discourage drug abuse. Protection might be used to protect the innocent from drug pushers. Government might need to enact more stringent drug prohibition laws. Recreation might be used to keep youth off the streets where drugs run rampant. Employment in the world of work might be used to reduce recidivism rates, and so on. Although health might be the area of the improvement sought, it would be foolhardy to limit the intervention to a medical or allied treatment only.

The Centers for Disease Control and Prevention was criticized in Chapter 3 for unnecessarily and counterproductively limiting and biasing its focus and efforts to improve quality of life. Consequently, all the money it spends to interview a million people to assess health-related quality of life in America may be relatively inconsequential, regardless of the spin the CDC puts on it in the news media.

The eight areas of concern were structured and presented in Table 3.5 to define quality of life as a field of study. The same format is presented in Table 7.1 to define the database in the field of study. The quality of life database format serves three purposes: (1) to indicate and assist in the formulation of

The Holistic Approach

scientific research and development to improve the quality of life, (2) to serve as a basis for making cogent decisions to improve the quality of life, and as mentioned (3) to circumscribe and define quality of life as a field of study or discipline. Continued research and development keeps the database current. This ensures that quality of life as a field of study will be alive and progressive and that the decisions based on the database will have the best chance of being effective, given the limits of current knowledge.

The format enables similarities as well as differences between societies and individuals to be analyzed. Deriving eight generic categories or areas allows societal and individual influences to be analyzed on a common basis. At the same time, differences in viewing quality of life from a societal and a personal point of view become evident.

The first challenge is to derive improvements to be pursued. It is not sufficient to identify quality of life indicators and attributes, as has been the practice. We need to aggressively attempt to improve quality of life. It is well to remember that, in general, professionals are implicitly committed to pursue particular improvements. For instance, doctors are implicitly committed to improve health. Educators are implicitly committed to improve learning. Desirable improvements can be identified by surveying professional literature. The survey will also disclose indicators and attributes that suggest desired improvements. Since the purpose is to establish a comprehensive list of improvements for each category, there is no need to be overly selective. However, since an improvement is defined as progress from an existing state to a desired state, both states need to be clearly defined. In addition, since the overall mission is to improve the quality of life, it should be made clear in the database how achieving a specified improvement can be expected to improve the quality of life.

It is important to ensure relevance and accuracy. Societal improvements are specified in normative terms that are derived from raw data. Existing and desired states are specified in such terms as *average, percentage*, or *rate*. On the other hand, existing and desired states defining personal improvements are specified as raw data, for example, qualitative and quantitative descriptions of characteristics of individuals. (See Tables 3.3 and 3.4 for examples). Treatments to bring about specific improvements are also described in professional literature. When an improvement can be readily attained, a specific treatment can be prescribed to achieve the improvement. In some cases, a number of specific treatments can be prescribed to achieve the improvement. In any event, all treatments that can be predicted to achieve an improvement should be listed in the database, with the success rate and possible side effects stipulated for each treatment along with treatment cost and availability and specifications. When an improvement cannot be readily attained, it is less likely that particular treatments can be specified with any certainty that they will be effective. In such cases, palliative and experimental treatments are described as such, and the need for research is highlighted.

Level of Concern

The levels that need to be considered in any effort to improve the quality of life are the societal level and the personal level. Although related, they are distinct, and it is necessary to distinguish between them to be effective. Attempts to improve quality of life in a society focus on making the environment a better place for people to live. Normative assessments are made to determine the improvements needed in a society or whether a particular locale is a good place to retire or locate a business. To rate the quality of life in a society, measures in all eight areas of concern might be needed, and the measures might be weighted to arrive at a rating for a particular purpose. The Index of Social Health described in Chapter 2 is an example of a procedure for assessing quality of life in a society.

Assessment of the quality of life of an individual is a different matter. Arguments were made in Chapter 2 that the quality of people's lives should be determined by obtaining their perceptions of the quality of their lives. We might simply ask them to rate the overall quality of their lives. Or we might ask them to rate the quality of different aspects of their lives in any or all of the eight areas of concern and derive a final QOL rating from the various ratings.

It is becoming increasingly evident that laws, such as those in Massachusetts, support sane individuals' rights to determine their best interests and ways to pursue them regardless of what objective evidence may demonstrate. And it can be legally incumbent on a professional to convey objective data to clients and to interpret the implications of the data to aid clients in coming to an informed decision. Figure 7.1 displays the structural factors discussed.

DYNAMIC FACTORS

Dynamics pertains to the cause-effect relations that generate improvement in quality of life. Relevant dynamics were introduced in Chapter 3, that is, the treatment → improvement relationship, in which improvement is the effect to be achieved and treatment is the causal agent used to achieve the effect. Another component of the dynamics that was fully explicated in Chapter 4 is the medium or vehicle that is used to develop treatments to achieve improvements, namely, the control cycle.

Still another component of the dynamics is the psychological forces that must be worked with to improve the quality of life: predictive ability and control. Predictive ability was discussed thoroughly in Chapter 6 as the mental factor largely responsible for superior human achievements. Control was discussed thoroughly in Chapter 5. It was shown that all people seek improved control of one thing or another, much of the time.

We will first consider the achievement of improvement, then treatments, and then the control cycle to further explicate the components of the dynamics.

The Holistic Approach

Figure 7.1
Structural Factors

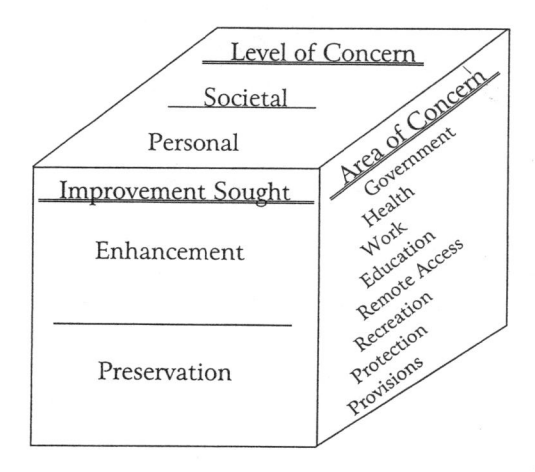

```
                    Level of Concern
                    Societal
                    Personal            Area of Concern
                    _____          Government
                    Improvement Sought   Health
                                         Work
                    Enhancement          Education
                                         Remote Access
                                         Recreation
                    _____          Protection
                                         Provisions
                    Preservation
```

Source: Author.

Predictive ability and control will be discussed as they are relevant to discussions of improvements, treatments, and the control cycle.

Achievement of Improvement

It has been shown that the improvement sought is most often increased control of movement from an existing state to a desired state, for example, increased control of movement from being pregnant to having a healthy baby. Although the focus of attention is usually on a particular improvement sought, the desire to control achievement of an improvement is almost always wanted as well, so that the improvement can be achieved at will in the future. The improvement sought is usually known and made explicit. The desire to control achievement of the improvement is implicit, and the people seeking the improvement are not necessarily aware of it at the time. However, it is more likely that an improvement will be achieved if the people who seek the improvement are aware that they need a control mechanism or treatment to effect the improvement. They are then better prepared to infer from an analysis of the pathway from the existing to the desired state the control mechanisms that are needed to effect the improvement.

The Treatment

Given any improvement sought, the objective is to develop a treatment that can be predicted to achieve the improvement without harmful side effects. In developing treatments, we tend to be concerned with the probability that a treatment will be effective with tolerable side effects without realizing that the purpose of computing probability is to estimate predictive ability. We preset a level of probability that represents the level of risk we are willing to take, and when a treatment is effective at that level of probability with tolerable side effects, we have attained the predictive ability we need to certify the use of the treatment.

Applying probability is the formal, mathematical way to establish predictive ability. And the estimation of probability is increasingly taught at higher levels of education. This is not surprising because it is used to predict success more often in an increasing number of enterprises, ranging from establishing insurance premiums that will yield a profit, to substituting a pinch-hitter in baseball, to polling the public to determine whether an aspiring politician has a chance of being elected.

What we neglect doing is developing the predictive ability of our youth progressing from the instruction they receive at home from their parents, continuing through elementary and high school, and culminating with the learning of statistics to estimate probability.

USING THE CONTROL CYCLE

The vehicle that is used to derive improvements and develop treatments that can be predicted to achieve the improvements is the control cycle. Figure 7.2 is a review of the control cycle.

Examples of personal and societal applications of the cycle were provided in Chapter 4. Here the important aspects of the cycle are highlighted, and the uses of the cycle from personal and societal perspectives are explicated and contrasted. Applications of the control cycle for societal purposes tend to be more formal, objective, and technically sophisticated. Personal applications tend to be more casual and are especially advantageous when people need to discipline themselves to pursue relatively long-term aspirations.

Projecting Improvements

Projecting improvements is more than goal setting. Goal setting is the projection of desired future states. Although it may be of some value to project goals initially as an expression of general aspirations, goal statements are an inadequate basis for planning treatments to fulfill aspirations or for determining the feasibility of administering the treatments. In projecting improvements, both the existing state, as a starting point, and the desired state, as a goal, are defined so that treatments can be planned to progress from the desired to the existing state, and the feasibility of administering the treatment can be roughly estimated beforehand. And as shown in Chapter 4, specifying both an existing and a desired state facilitates determining the need for improvement initially, deriving

Figure 7.2
The Control Cycle

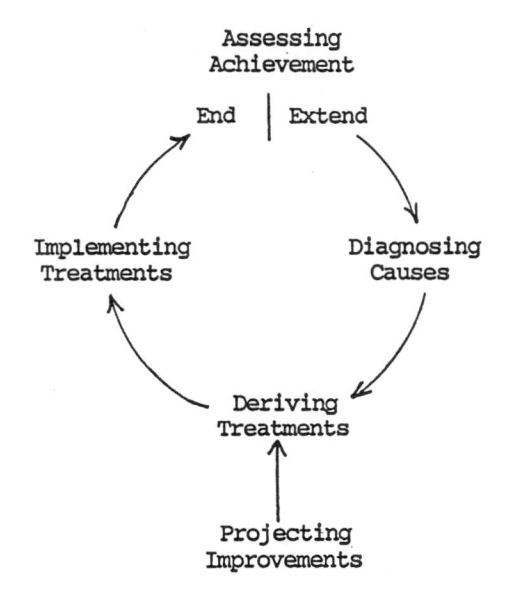

```
                    Assessing
                   Achievement
              End  |  Extend
                ↗     ↘
    Implementing         Diagnosing
     Treatments            Causes
           ↖             ↙
                Deriving
               Treatments
                   ↑
                   |
               Projecting
              Improvements
```

Source: Author.

treatments to effect improvements that are needed, and assessing the achievement of improvements.

The need for an improvement is manifested as a discrepancy between an existing and a desired state. In defining discrepancies, sometimes the existing state is noted and then the desired state is projected. Such is the case when people become aware of an imminent danger and project a future state that brings relief. At other times the desired state is specified first as a goal or aspiration. Then the existing status is assessed to determine the extent of the discrepancy, if any. At times, it is found that there is no need for a contemplated improvement. At other times it may become evident that achievement of the desired state is unrealistic, given the existing state.

Societal Applications. Societal applications pertain to the achievement of public improvements by government and private enterprise. Government agencies tend to pursue improvements that lobbyists and the public are pressing to achieve. Private enterprises pursue improvements that are profitable to achieve. Sometimes government agencies become involved in pursuing improvements that are not profitable to achieve, for the public good.

Several kinds of research and development techniques would aid in determining the need for improvement. *Policy research* often needs to be conducted to achieve consensus among policy makers on a desired state to pursue. The importance of determining the desired state authoritatively and objectively by qualified people was stressed.

Instrument development may need to be undertaken to observe the existing state. An accurate observation procedure is needed to determine whether improvement is needed in the first place and to assess achievement of improvement. Before an observation instrument is ready for use, it must be established that the instrument yields valid, reliable, and objective observations. Moreover, it is of great importance that the instrument have diagnostic capability so that causes of inadequate improvement can be diagnosed. *Evaluative research* also may need to be conducted to compare the existing and desired states in order to determine whether a discrepancy exists, indicating a need for improvement.

Statistics need to be used in policy research to determine whether the agreement among policy makers is sufficient and statistically significant. This typically requires a test of correlation. Statistics need to be used in instrument development to establish the validity, reliability, and objectivity of the instruments being constructed. Objectivity and reliability are established by means of correlation. Validity may require tests of either correlations or differences, depending on the type of validation being sought. Evaluation research requires tests of differences to determine whether there is a statistically significant discrepancy between the desired and the existing states.

Personal Applications. Although individuals can avail themselves of objective evidence and research tools such as statistics to come to a personal decision, the use of objective evidence and research tools requires more education than most people have been afforded in the past. Knowing how to use objective evidence in decision making is a requirement for having an informed public. And many sophisticated people use objective evidence to make decisions while shopping. For instance, they examine labels on food products to compare the cost of food items per ounce, calories, and cholesterol levels. On the other hand, knowing how to conduct research may only be necessary for researchers

From a personal perspective, projecting improvements is largely a matter of expressing one's aspirations (the desired state) and then getting a general idea of the feasibility of fulfilling one's aspirations, given one's present circumstances and available resources (the existing state).

Deriving Treatments

Treatments are derived to achieve improvements. If an intervention is called for, the following format can be followed to derive treatments. The format was exemplified in Chapter 6.

Determining the factors to be controlled. First, the discrepancy defining the improvement to be achieved is analyzed to determine the factors that need to be controlled in order to achieve the improvement. The question that needs to be answered is, What factors need to be controlled to progress from the existing to the desired state? The question is answered by consulting theory and the research literature to identify causal agents that can effect the improvement.

Determining constraints. This amounts to identifying factors that must be worked with and worked around in achieving the improvement, including resource limitations and possible harmful side effects.

Determining means of controlling the factors. Means of manipulating the causal agents that can effect the improvement are identified from a review of relevant theory and research literature. This amounts to identifying as many ways as possible of manipulating each causal agent identified to effect the improvement.

Deriving a treatment. Means of manipulating the causal agents to effect the improvement are selected, and a procedure is written specifying how, when, and where the causal agents are manipulated to effect the improvement.

Innovative treatments are derived when a new way or a new combination of ways of manipulating a causal agent is identified and tested. The potential for deriving an innovative treatment to effect an improvement is increased each time a new causal agent is linked with the achievement of an improvement or a new way of manipulating a causal agent as desired is developed.

Societal Applications. The general mission of societal enterprises, government and private enterprises alike, is to provide treatments for the public that can be predicted to effect improvements without detrimental side effects. To endure, private enterprises must develop treatments that are profitable. Governments subsidize the development of treatments that are not profitable to develop, for instance, treatments for rare, pernicious diseases. Governments also are responsible for certifying the effectiveness and safety of treatments to be used by the public.

To be productive in the development of effective treatments, societal research and development needs to proceed systematically. *Experimental research* needs to be conducted to test the effectiveness of treatments that can be hypothesized to bring about needed improvements. When potentially effective treatments are not available to be tested, *treatment development* needs to be engaged in to produce such treatments. When the knowledge base is not sufficient to develop potentially effective treatments, *causal-comparative research* needs to be conducted to identify causal agents that might be manipulated to induce a needed improvement. Once causal agents are identified, treatments can be developed to manipulate the causal agents to induce the improvement. When treatments are developed to induce the improvement, experimental research can be employed to test the effectiveness of the treatments. Although predictive research and descriptive research contribute to treatment development by increasing knowledge about possible causal agents, this only indirectly promotes the development of effective treatments.

Present efforts to keep the public informed and enlightened on how to improve the quality of their lives leave a great deal to be desired. In many respects the public is not provided with sufficient advice to enable them to help themselves. The Food and Drug Administration is advocating that people be given easy-to-understand information on how to take a prescribed medicine and what side effects to anticipate. The nation spends $20 billion a year treating side effects and illnesses from improper use of medicines, the government says (1995). In other respects, people are provided with too much specialized detail on particular ailments, more than they can assimilate and more than they need to know to improve the quality of their lives. In this day of proliferating specialization, an increasing volume of information is being published in an increasing number of health and allied professions. Even when specialists such as doctors attempt to publish newsletters for the public, they provide an excessive amount of specialized detail, often in the jargon of their profession. And they do not relate the contribution of their advice to the overall quality of people's lives. What's needed is the presentation to the public of fundamental, holistic information on the various factors essential to quality of life. The public needs to know how the eight factors in the proposed guidelines can contribute to the quality of their lives (government, health, work, education, remote access, protection, provisions, recreation). Perhaps the public is ready for a quality of life publication that offers them advice on how to improve the quality of life taking all eight factors into account.

Personal Applications. Treating oneself has always been popular, especially when it involves preventing loss of control that can compromise the quality of life. Although there is an overabundance of ads that recommend treatments for maintaining health, the regimens people adopt for themselves and recommend to their friends are most often idiosyncratic. People tend to add their own variations to a theme. They might consult nutritionists before deciding on dietary practices and supplements. But their total diet will usually consist of their own combination of foods, vitamins, minerals, herbs, and elixirs. Similarly, many people concoct their own exercise regimens even if they join health clubs and are given professional advice.

With the vast number of over-the-counter drugs available, more and more people are inclined to treat themselves for ailments. They try to diagnose their symptoms and peruse pharmacies, reading labels on over-the-counter drugs until they find one they think will relieve their symptoms. Or they might choose one a trusted friend or relative recommends. Of course, many will seek and take their druggists' advice. Ultimately, however, they decide which over-the-counter medication they will take.

Resourceful people also develop ways to relieve their mental stress that inevitably accrues from the pressures of daily living and from occasional crises. Talking to people who are empathetic, sympathetic, and understanding is always helpful, whether one finds them at a neighborhood bar, at work, or in the neighborhood where one lives or they happen to be lifelong friends. More formal conversations with clergymen, counselors, or psychotherapists are at least as helpful. However, taking calming drugs can backfire. Something as ostensibly innocuous as over-the-counter sedatives, pain relievers, sleeping pills, and antihistamines can seriously impede mental and physical dexterity. People can also relieve their tension through sports, entertainment, recreation, socializing, and spiritual solace.

Finally, people must be able to find and engage capable assistance when they are unable to help themselves. Whether they need advice or treatment, too many people do not know how to avail themselves of competent professional treatment.

In addition to being instructed on how to evaluate objective evidence to decide on a diet or exercise regimen, an over-the-counter drug, or a form of relaxation therapy, people need to be taught how to access competent professional assistance.

Implementing Treatments

In order to test the effectiveness of a treatment, it is necessary to execute it accurately. It is not uncommon for the implementation of a treatment to be invalid because deviations from the treatment plan were too pronounced. So every effort needs to be made to execute a treatment accurately, especially complicated and protracted treatments. Treatments need to be learned and mastered before they are executed. The mastery of a complicated treatment often requires extended training rehearsal and practice.

Societal Applications. Learning to master the execution of treatments in business, government, and educational enterprises requires formal attention. In addition to memorizing treatment specifications, the execution of the treatment needs to be practiced until it is perfected. Before a treatment is applied, (1) the execution of the treatment must be observed, (2) the actual execution must be compared to the treatment specifications, and (3) discrepancies between the two must be corrected until the execution of the treatment can be certified as accurate. Afterward, while the treatment is being applied, its execution must be observed and discrepancies must be corrected before they become too pronounced. Video and audio recordings of the execution of the treatment are frequently helpful in finding and correcting discrepancies.

Instructors who teach people how to execute treatments must be trained to assess people's knowledge of the treatment specifications before they attempt to execute the treatment and to detect and correct discrepancies between actual execution and treatment specifications.

Implementing treatments accurately requires *evaluative research* to detect differences between treatment specifications and treatment administration.

Personal Applications. The treatments people implement on their own range from taking necessary medication precisely as prescribed to the casual execution of recipes that might be altered to taste to the sober application of recommended child-rearing practices. Most often, these applications require little in the way of extended formal training. However, administering some medical treatments to oneself may require training, and parents might take classes and workshops on child rearing. There are occasions on which people need to arrange for the detection and correction of deviations from the prescribed behaviors—for example, when they are memorizing lines for a play or a speech.

Assessing Achievement

The purpose of assessing achievement is to determine, after the treatment has been completed, the extent to which progress has been made, as well as the side effects of the treatment. This requires that the actual outcome be observed; that the actual outcome be compared to the desired state to determine the progress that has been made, if any; that the relationship between the actual outcome and the desired state be interpreted; and that the treatment recipients be contacted and observed to uncover side effects. Fundamentally, one of three decisions can be made as a result of assessing achievement: (1) If the desired state is reached, the pursuit ends. If the desired state is not reached, the pursuit of the desired state either (2) extends to the diagnosis of causes or (3) is aborted for other reasons, for instance, inadequate resources to continue or a change in priorities.

Societal Applications. Assessing achievement in business, government, and education requires *evaluative research*—and then some. Methods of observing the actual outcome accurately need to be available. This might necessitate the use of a scale of measurement and/or an observation instrument such as a test or a video or audio recorder. Statistics may need to be used to formally compare the outcome to the desired state to determine whether any discrepancy may be a chance factor. This, too, is evaluative research. The interpretation of the findings can be an easy or overwhelming job. If the assessment is too extensive or complex, the findings may be too variable to infer a straightforward conclusion. The purpose of the assessment is to test the hypothesis that the treatment will effect the achievement of the desired state, and it should be kept simple and focused on testing that hypothesis. Furthermore, the bias of the interpreter needs to be guarded against, by arranging beforehand for an objective interpretation of the findings. To assess side effects accurately, every treatment recipient must be systematically interviewed and observed. It is insufficient and often misleading to rely solely on voluntary recipient complaints.

Personal Applications. Personal assessment of achievement is very much subject to subjective interpretation. Regardless of how much objective evidence may be available, the interpretation of treatment effectiveness depends on people's reaction to the treatment: the impact the treatment may have had on them, the cost of the treatment to them, the time it took to administer the treatment, the relief the treatment gave them, and the side effects of the treatment. Treatments have idiosyncratic effects on people, and their assessment of treatment effectiveness will result from their reactions to the many idiosyncratic effects the treatment may have had on them. However, despite idiosyncratic personal opinions, agreement can be found among recipients on many treatment effects.

Diagnosing Causes

When pursuit of an improvement is extended, the causes of the actual outcome are diagnosed so that enlightened modifications can be made to the treatment in order to achieve the improvement on a subsequent attempt. Diagnosing causes entails reconstructing the antecedent conditions that led to the outcome and determining from analysis of the antecedent conditions the causes of the outcome. The process then proceeds once again to the next stage, "Deriving Treatments," where a new treatment is derived, based on the diagnosis of causes, that has a better chance of achieving the improvement being pursued.

Societal Applications. Instrumentation is often developed to enhance formal diagnostic procedures in business, government, and education. Some instruments are developed to make direct observations to enhance the diagnosing of causes. For example, cameras are used in banks to monitor traffic so that culprits may be identified after a bank robbery. Other instruments are developed to make indirect observations, for example, paper-and-pencil diagnostic instruments. In Chapter 4 the importance of using instruments to diagnose causes of failure on the job was emphasized to enable evaluators who are assessing people's achievements to recommend ways for people to improve their performances. Without diagnostic information, there is no basis for advising people on how to correct their mistakes.

Personal Applications. Personal applications of diagnosing causes often pertain to the foibles of daily living. We diagnose causes when we can't find an item we mislaid. If we are accurate in our reconstruction of our prior activities, we are able to remember where the item was left. Personal applications also pertain to recreational activities such as reading who-done-its. We try to construct the cause of the crime before the author reveals it.

It may be helpful to end the book showing the parts/whole relationships of the holistic approach in topic outline form:

I. Structural factors
   A. The improvement sought
      1. Preservation
      2. Enhancement
   B. Areas of concern
      1. Government
      2. Health
      3. Work
      4. Education
      5. Remote access
      6. Recreation

       7. Protection
       8. Provisions
   C. Level of concern
       1. Societal
       2. Personal
II. Dynamic factors
   A. Effect sought: achievement of improvement
   B. Causal agent: treatment
   C. Vehicle: the control cycle
       1. Projecting improvements
       2. Deriving treatments
       3. Implementing treatments
       4. Assessing achievement
       5. Diagnosing causes
   D. Psychological forces
       1. Predictive ability: the mental ability largely responsible for achievement
       2. Improved control: an achievement all people seek much of the time

References

Acton, R.G., and During, S. 1990. The treatment of aggressive parents: An outline of a group treatment program. *Canada's Mental Health* 38: 2–6.
Adams, J.A., and Xhigriesse, L.V. 1960. Some determinants of two-dimensional tracking behavior. *Journal of Experimental Psychology* 60: 391–403.
Aristotle. 1934. *The Nicomachean ethics*. The Loeb Classical Library. Cambridge, MA: Harvard University Press.
Baldwin, S., and Gerard, K. 1990. Caring at home for children with mental handicaps. In *Quality of Life*, ed. S. Baldwin, C. Godfrey, and C. Propper. London: Routledge.
Bales, R.F. 1950. *Interaction process analysis: A method for the study of small groups*. Reading, MA: Addison-Wesley.
Baltes, M., and Baltes, P., eds. 1986. *The psychology of control and aging*. Hillsdale, NJ: Lawrence Erlbaum.
Banziger, G., and Roush, S. 1983. Nursing home for the birds: A control relevant intervention with bird feeders. *Gerontologists* 23(5) (October): 527–531.
Benz, D., and Rosemier, R. 1966. Concurrent validity of the Gates level of comprehension test and the Bond, Clymer, Hoyt reading diagnostic tests. *Educational and Psychological Measurement* 26: 1057–62.
Best places to live in America. 1994. *Money* 23(9) (September): 127–142.
Binding, K., and Hoche, A. 1975. *The release of the destruction of life devoid of value*. Ed. R. Sassone. Santa Anna, CA: Life Quality Paperbacks.
Bindra, D.A. 1959. *Motivation*. New York: Ronald Press.
Birren, J.E., Lubben, J.E., Rowe, J.C., and Deutchman, D.D. 1991. *The concept and measurement of quality of life in the frail elderly*. San Diego, CA: Academic Press.
Blau, T.H. 1977. Quality of life, social indicators, and predictors of change. *Professional Psychology* 8: 464–473.

Bloch, J.H., and Anderson, L.W. 1975. *Mastery learning in classroom instruction.* New York: Macmillan.
Bloom, B.S. 1968. Learning for mastery. *Evaluation comment* 1: 1–11.
Brickman, P., and D'Amato, B. 1975. Exposure effects in a free-choice situation. *Journal of Personality and Social Psychology* 32: 415–420.
Brill, P.A., Burkhalter, H.E., Kohl, H.W., and Blair, S.N. 1989. The impact of previous athleticism on exercise habits, physical fitness, and coronary heart disease risk factors in middle aged men. *Research Quarterly for Exercise and Sport* 60(3) (September): 209–214.
Brock, D. 1993. Quality of life measures in health care and medical ethics. In *The quality of life*, ed. M. Nussbaum and A. Sen. Oxford, England: Clarendon Press.
Brown, L.H. 1988. Dropping out: From prediction to prevention. Paper presented at the annual meeting of the American Educational Research Association, New Orleans, LA, April 5–9.
Campbell, A. 1976. Subjective measures of well-being. *American Psychologist* (February): 117–124.
Campbell, A., Converse, P.E., and Rodgers, W.L. 1976. *The quality of American life: Perceptions, evaluations, and satisfactions.* New York: Russell Sage Foundation.
Carroll, K.M., Rounsaville, B.J., and Keller, D.S. 1991. Relapse prevention strategies for the treatment of cocaine abuse. *American Journal of Drug and Alcohol Abuse* 17: 249–265.
Centers for Disease Control and Prevention. 1995. Health related quality of life measures. *Morbidity and Mortality Weekly Report* 44(11) (March 24): 195–200.
Centers for Disease Control and Prevention. 1994. Measuring health-related quality of life for public health surveillance. *Morbidity and Mortality Weekly Report* 43(20) (May 27): 375–380.
Centers for Disease Control and Prevention. 1991. Workshop on quality of life/health status surveillance for states and communities. Stone Mountain, GA, December 2–4, meeting report.
Charmello, C. 1993. Self-questioning prediction strategy's effect on comprehension. Master's thesis, Kean College of New Jersey.
Chia, T. 1995. Learning difficulty in applying notion of vector in physics among "A" level students in Singapore. [ED389528]
Chick, T.W., Cagle, T.G., Vegas, F.A., Poliner, J.K., and Murata, G.A. 1991. The effects of aging on submaximal exercise performance and recovery. *Journal of Gerontology* 46(1): 34–38.
Chubon, R.A. 1990. *Life situations survey test manual.* Columbia: University of South Carolina.
Consumer Reports on health. 1995. *Consumer Reports* (April).
Corden, A. 1990. Choice and self-determination as aspects of quality of life in private sector humans. In *Quality of life*, ed. S. Baldwin, C. Godfrey, and C. Propper. London: Routledge.
Darwin, Sir F. 1950. *Charles Darwin's autobiography.* With notes and letters deciphering the growth of *The origin of the species.* New York: Henry Schuman.
DeCharms, R. 1968. *Personal causation: The internal affective determinants of behavior.* New York: Academic Press.
Delbelq, A.L., Van deVen, A.H., and Gustafson, D.H. 1975. *Group techniques for pro-*

gram planning: A guide to nominal group and Delphi processors. Glenview, IL: Scott, Foresman.

Diwan, S. 1994. *Unmet health needs and quality of life of the elderly*. Development of a survey for states and communities (technical report). Atlanta, GA: Centers for Disease Control and Prevention, Aging Studies Branch.

Doherty, W. 1983. Locus of control and marital interaction. In *Research with the Locus of Control Construct, Volume 2: Developments and Social Problems*, ed. H.M. Lefcourt. New York: Academic Press.

Donovan, D., and O'Leary, M. 1983. Control orientation, drinking behavior, and alcoholism. In *Research with the Locus of Control Construct, Volume 2: Developments and Social Problems*, ed. H.M. Lefcourt. New York: Academic Press.

Downie, R.S., Fyfe, C., and Tannahill, A. 1990. *Health promotion: Models and values*. New York: Oxford University Press.

Dustman, R.E., Ruhling, R.O., Russell, E.M., Shearer, D.E., Bonekat, W., Shigeoka, J.W., Wood, J.S., and Bradford, D.C. 1984. Aerobic exercise training and improved neurological function of older individuals. *Neurology of Aging* 5: 35–42.

Dykes, S. 1997. A test of proposition, of prediction theory. Doctoral dissertation, University of South Carolina, Columbia, SC.

The Encyclopedia of Magic and Superstition. 1974. London: Octopus Books Ltd. and Phoebus Publishing Co.

Erickson, P., Wilson, R., and Shannon, I. 1995. Years of healthy life: Charting improvements in the nation's health. In *Quality of life in behavioral medicine*, ed. J. Dimsdale and A. Baum. Hillsdale, NJ: Lawrence Erlbaum.

Erickson, R. 1993. Descriptions of inequality: The Swedish approach to welfare research. In *The quality of life*, ed. M. Nussbaum and A. Sen. Oxford, England: Clarendon Press.

Etscheidt, S. 1991. Reducing aggressive behavior and improving self-control: A cognitive-behavioral training program for behaviorally disordered adolescents. *Behavioral Disorders* 16: 107–115.

Fillenbaum, G.G. 1988. *Multidimensional functional assessment of older adults: The Duke, older Americans, resources and services*. Hillsdale, NJ: Lawrence Erlbaum.

Findley, M., and Cooper, H. 1983. Locus of control and academic achievement: A literature review. *Journal of Personality and Social Psychology* 44(2): 419–427.

Fisher, J. 1986. Maggie Kuhn's vision: Young and old together. *50 Plus* 26 (July): 22–24.

Flanagan, J. 1978. A research approach to improving the quality of life. *American Psychologist* (February): 138–147.

Foreyt, J.P., and Goodrick, G.K. 1991. Factors common to successful therapy for the obese patient. *Medicine and Science in Sports and Exercise* 23: 292–297.

Freed, M.M. 1984. Quality of life: The physician's dilemma. Presidential address. *Archives of the American Academy of Physical Medicine and Rehabilitation* 65 (March).

Freeman, R. 1982. Improving the comprehension of stories using predictive strategies. Paper presented at the annual meeting of the International Reading Association.

Friedman, M.I. 1993. *Taking control: Vitalizing education*. Westport, CT: Praeger.

Friedman, M.I. 1974. *Predictive ability test*. Manual published by author.

Friedman, M.I., and Lackey, G.H., Jr. 1991. *The psychology of human control: A general theory of purposeful behavior*. Westport, CT: Praeger.

Friedman, M.I., and Willis, M.R. 1981. *Human nature and predictability*. Lexington, MA: Lexington Books.
Fromm, E. 1971. *Escape from freedom*. New York: Avon Books.
Gelber, R.D., Gelman, R.S., and Goldhirsch, A. 1989. A quality of life oriented endpoint for comparing therapies. *Biometrics* 45: 781–795.
George, L.K., and Bearon, L.B. 1980. *Quality of life in older persons: Meaning and measurement*. New York: Human Sciences Press.
Gibbs, J.P. 1989. *Control: Sociology's central notion*. Urbana and Chicago: University of Illinois Press.
Gill, T.M., and Feinstein, A.R. 1994. A critical appraisal of the quality of quality-of-life measurements. *Journal of the American Medical Association* 272: 619–626.
Glasser, W. 1984. *Take effective control of your life*. New York: Harper and Row.
Gordon, D. 1977. Children's beliefs in internal-external and self-esteem as related to academic achievement. *Journal of Personality Assessment* 41(4): 333–336.
Gorovitz, S. 1982. *Doctor's dilemma: Moral conflict and medical care*. New York: Macmillan.
Gray power. 1990. *The Nation* 250 (May): 727.
Greene, J., and Noreen, D. 1974. Time to read semantically related sentences. *Memory and Cognition* 2: 117–120.
Guadagnoli, E. 1988. Quality of life measurement: A psychometric tower of Babel. *Journal of Clinical Epidemiology* 41(11): 1055–1058.
Guyatt, G., and Cook, D. 1994. Health status, quality of life, and the individual. *Journal of the American Medical Association* 272(8): 630–631.
Harish v. Children's Hospital Medical Center. 1982. 38 Mass. 152.
Hart, B.A. 1981. The effects of age and habitual activity on the fractionated components of resisted and unresisted response time. *Medicine and Science in Sports and Exercise* 13: 78.
Hayden, B., Nasby, W., and Davids, A. 1977. Interpersonal conceptual structure, predictive accuracy and social adjustment of emotionally disturbed boys. *Journal of Abnormal Psychology* 86: 315–320.
Healthy People 2000. 1991. Washington, DC: U.S. Government Printing Office.
Healy, R. 1963. Does preoperative instruction make a difference? *American Journal of Nursing* 68(1): 62–67.
Heller, J. 1961. *Catch-22*. New York: Simon & Schuster.
Henderson, E., and Long, B. 1968. Correlation of reading readiness and children of varying backgrounds. *The Reading Teacher* 22: 40–44.
Hennessey, C.H., Moriarty, D.G., Zack, M.M., Schear, P.A., and Brackbill, A. 1994. Measuring health-related quality of life for public health surveillance. *Public Health Reports* (Centers for Disease Control and Prevention, Atlanta, GA) 109 (5): 665–672.
Hickson, J., Housely, W.F., and Boyle, C. 1988. The relationship of locus of control to life satisfaction and death anxiety in older persons. *International Journal of Aging and Human Development* 26(3).
Hodgson, J.L., and Buskirk, E.R. 1977. Physical fitness and age, with emphasis on cardiovascular function in the elderly. *Journal of the American Geriatric Society* 25: 385–392.
Home guide. 1994. *U.S. News and World Report* (April 11): 57–83.

References

Hornquist, J.O. 1982. The concept of quality of life. *Scandinavian Journal of Social Medicine* 10: 57–61.
Hunt, J., and D. Joseph. 1990. Using prediction to improve reading comprehension of low-achieving readers. *Journal of Clinical Reading, Research and Programs* 3(2): 14–17.
Hurst, R., and Milkent, M. 1994. Facilitating successful predictive reasoning in biology through application of skill theory. Paper presented and the annual meeting of the National Association for Research in Science Teaching, Anaheim, CA, March 19–26. [ED368582]
Ibsen, H. 1935. *Eleven plays by Henrik Ibsen.* New York: Modern Library.
Johns Hopkins Medical Newsletter. 1994. *Health after 50.* Baltimore, MD.
Kaplan, R.M., and Bush, J.W. 1982. Health-related quality of life measurement for evaluation, research and policy analysis. *Health Psychology* 1: 61–80.
Kasch, F.W., Boyer, J.L., Van Camp, S., Verity, L.S., and Wallace, S.P. 1990. The effect of physical activity and inactivity on aerobic power in older men. *Physician and Sports Medicine* (April 18): 73–83.
Kelly, G.A. 1963. *A theory of personality: The psychology of personal constructs.* New York: W.W. Norton.
Kelly, G.A. 1955. *The psychology of personal constructs.* Vol. 1. New York: W.W. Norton.
Kind, P., Claire, G., and Godfrey, C. 1990. What are Qualys? In *Quality of life: Perspectives and policies*, ed. S. Baldwin, C. Godfrey, and C. Propper. London: Routledge.
Kohler, W. 1969. *The task of Gestalt psychology.* Princeton: Princeton University Press.
Kohler, W. 1929. *Gestalt psychology.* New York: Leveright.
Kohler, W. 1925. *The mentality of apes.* New York: Harcourt.
Kuchkremer, G., Milnneker, E., and Block, M. 1991. Smoking cessation treatment combining transdermal nicotine substitution with behavioral therapy. *Pharmacopsychiatry* 24: 96–102.
Lamb, J. 1984. *A fine age: Creativity as a key to successful aging.* Little Rock, AR: August House Publishers.
Langer, E. 1983. *The Psychology of Control.* Beverly Hills, CA: Sage Publications.
Langer, E., and Rodin, J. 1976. The effects of choice and enhanced personal responsibility for the aged: A field experiment in an institutional setting. *Journal of Personality and Social Psychology* 34(2) (August): 191–198.
Langer, E., Rodin, J., Beck, P., Weinman, C., and Spitzer, L. 1979. Environmental determinants of memory improvement in late adulthood. *Journal of Personality and Social Psychology* 37 (November): 2003–2013.
Lemke, S., and Moos, R. 1981. The suprapersonal environments of sheltered care settings. *Journal of Gerontology* 36(2): 233–243.
Leonard, J.A. 1953. Advance information in sensory-motor skills. *Quarterly Journal of Experimental Psychology* 5: 141–149.
Lesieur, H.R., and Rosenthal, R.J. 1991. Pathological gambling: A review of the literature. *Journal of Gambling Studies* 7: 5–39.
London, H., Schubert, D.S., and Washburn, D. 1972. Increase in automatic arousal by boredom. *Journal of Abnormal Psychology* 80: 29–36.
Lyman, F. 1988. Maggie Kohn: A wrinkled radical's crusade. *The Progressive* 52 (January): 29–31.

Manson, J., Willett, W., Stampfer, M., Colditz, G., Hunter, D., Hankinson, S., Hennekens, C., and Speizer, F. 1995. Body weight and mortality among women. *The New England Journal of Medicine* 333(11): 677–685.
May, R., and Hutt, C. 1974. Response to stimulus uncertainty in four-, six-, and eight-year-old children. *Journal of Psychology* 88: 127–133.
McMurran, M. 1991. Young offenders and alcohol related crime: What interventions will address the issue? *Journal of Adolescence* 14: 245–253.
Megone, C. 1990. The quality of life starting from Aristotle. In *Quality of life: Perspectives and policies*, ed. S. Baldwin, C. Godfrey, and C. Propper. London: Routledge.
Menotti, F.S. 1992. Physical activity, physical fitness, and mortality in a sample of middle-aged men followed-up 25 years. *Journal of Sports Medicine and Physical Fitness* 32(2) (June): 206–212.
Mezey, A. 1986. *People Weekly* (February 24): 93.
Michener, J.A. 1970. *The quality of life*. Philadelphia: J.B. Lippincott.
Miringoff, M.L. 1995. *Index of social health*. Tarrytown, NY: Institute for Innovation in Social Policy, Fordham University.
Moore, D.H. 1975. A study of age group track and field records to relate age and running speed. *Nature* 253: 264–265.
Moos, R. 1981. Environmental choice and control in community care settings for older people. *Journal of Applied Social Psychology* 11(1) (January-February): 23–43.
Moos, R., and Ingra, A. 1980. Detriments of the social environments of sheltered care settings. *Journal of Health and Social Behavior* 21: 88–98.
Moritani, T., and deVries, H.A. 1980. Potential for gross muscle atrophy in older men. *Journal of Gerontology* 35: 672–682.
National Center for Statistics. 1995. Years of healthy life, by P. Erickson, R. Wilson, and I. Shannon. *Statistical Notes* (7) (April): 1–14.
National Institutes of Health. 1993. Quality of life assessment. Proceedings of a workshop. Bethesda, MD: U.S. Department of Health and Human Services.
National Institutes of Health. 1990. Quality of life assessment. Proceedings of a workshop, October 15–17. Bethesda, MD: U.S. Department of Health and Human Services.
Natriello, G. 1987. *School dropouts: Patterns and policies*. New York: Teachers College Press.
Neisser, U. (Task Force chair). 1996. Intelligence: Knowns and unknowns. *American Psychologist* 51(2): 77–101.
Nobody took me seriously. 1993. *The State Newspaper, Parade Magazine* (Columbia, SC), insert, May 23.
Nolan, T. 1991. Self-questioning and prediction: Combining metacognitive strategies. *Journal of Reading* 35(2): 132–138.
Nordenfeldt, L. 1993. *Quality of life: Health and happiness*. Brookfield, VT: Avebury.
Nussbaum, M. 1993. Non-relative virtues: An Aristotelian approach. In *Quality of life*, ed. M. Nussbaum, and A. Sen. Oxford, England: Clarendon Press.
O'Brien, S.J., and Vertinsky, P.A. 1991. Unfit survivors: Exercise as a resource for aging women. *Gerontologist* 31: 347–352.
O'Leary, A. 1985. Self-efficacy and health. *Journal of Behavior Research and Therapy* 23: 437–451.
Ollendick, T.H., Hagopian, L.P., and Huntzinger, R.M. 1991. Cognitive behavior therapy

with nightime fearful children. *Journal of Behavior Therapy and Experimental Psychiatry* 22: 113–121.
Ostrow, A. 1984. *Physical activity in the older adult*. Princeton, NJ: Princeton Book Company.
Padilla, G.V., Presant, C., Grant, M.M., Matter, G., Lipsett, J., and Heide, F. 1983. Quality of life index for patients with cancer. *Research in Nursing and Health* 6: 117–126.
Paffenberger, R.S., Hyde, R.T., Wing, A.L., and Hsieh, C.C. 1986. Physical activity, all cause mortality and longevity of college alumni. *New England Journal of Medicine* 314: 605–613.
Parker, G. 1990. Spouse careers: Whose quality of life? In *Quality of life*, ed. S. Baldwin, C. Godfrey, and C. Propper. London: Routledge.
Patrick, D.L., and Erickson, P. 1993. *Health status and health policy: Quality of life in health care evaluation and resource allocation*. New York: Oxford University Press.
Pervin, L. 1963. The need to control and predict under conditions of threat. *Journal of Personality* 31: 570–587.
Petoski, J. 1991. Late bloomer. *Texas Monthly* 19 (October): 116.
Pfeiffer, E., ed. 1975. *Multidimensional functional assessment: The Oars methodology*. Durham, NC: Center for the Study of Aging and Human Development, Duke University Medical Center.
Powers, W. 1973a. *Behavior: The control of perception*. Chicago: Aldine.
Powers, W. 1973b. Feedback: Beyond behaviorism. *Science* 179: 351–356.
Proctor, R.N. 1988. *Racial hygiene: Medicine under the Nazis*. Cambridge, MA: Harvard University Press.
Pumping iron helps granny too. 1990. *Science News* 137 (June 23): 398.
Reutzel, D., and Fawson, P. 1991. Literature webbing predictable books: A prediction strategy that helps below-average, first-grade children. *Reading Research and Instruction* 30(4): 20–30.
Robine, J.M., and Branch, L.G. 1992. Measurement and utilization of healthy life expectancy: Conceptual issues. *Bulletin of the World Health Organization* 70(6): 791–800.
Rodin, J., and Langer, E.J. 1977. Long-term effects of control-relevant intervention with the institutionalized aged. *Journal of Personality and Social Psychology* 35(12): 897–902.
Rontoyannis, G.P. 1992. Sixty-three years of competitive sport activity. *Journal of Sports Medicine and Physical Fitness* 2(3) (September): 332–339.
Rosser, R., and Kind, P. 1978. A scale of valuations of states of illness: Is there a social consensus? *International Journal of Epidemiology* 7(4): 347–358.
Rothlein, L. 1988. Innerviews. *Women's Sports and Fitness* 10 (June–July): 64.
Rotter, J.B. 1975. Some problems and misconceptions related to the construct of internal versus external control of reinforcement. *Journal of Consulting and Clinical Psychology* 43: 56–67.
Rotter, J.B. 1966. Generalized expectancies for internal versus external locus of control of reinforcements. *Psychological Monographs* 180(1) (whole no. 609).
Rumberger, R.W. 1987. High school dropouts: A review of issues and evidence. *Review of Educational Research* 57: 101–121.
Rusk, H.A. 1972. *A world to care for*. New York: Random House.

Rusk, H.A. 1964. Preventive medicine, curative medicine then rehabilitation. *New Physical* 13: 165–167.

Sandvik, L., Erikssen, J., Thanlow, E., Erikssen, G., Mundal, R., and Rodahl, K. 1993. Physical fitness as a predictor of mortality among healthy middle-aged Norwegian men. *New England Journal of Medicine* 238(8) (February 25): 533–537.

Savitz, J.C. 1979. Diagnosis and treatment of emotionally disturbed clients using rational behavior theory. Ph.D. diss., University of South Carolina.

Schalock, R.L. 1990. *Quality of life: Perspectives and issues.* Washington, D.C.: American Association on Mental Retardation, 141–148.

Scheer, L. 1980. Experience with quality of life comparisons. In *The quality of life*, ed. F. Szalai and and F.M. Andrews. Beverly Hills, CA: Sage Publications.

Schneider, M. 1976. The quality of life and social indicators. *Public Administration Review* (May–June).

Schulz, R., and Heckhausen, J. 1996. A life span model of successful aging. *American Psychologist* 51(7): 702–714.

Schultz, N.R., Jr., and Hoyer, W.J. 1976. Feedback effects on spatial egocentrism in old age. *Journal of Gerontology* 31(1) (January): 72–75.

Scitovsky, T. 1976. *The joyless economy.* New York: Oxford University Press.

Shakespeare, W. 1936. A comedy of errors. In *The Complete Works of William Shakespeare*, ed. W. Wright. Garden, City, NY: Garden City Books.

Sherwood, D.E., and Selder, D.S. 1979. Cardiovascular health, reaction time, and aging. *Medicine and Science in Exercise and Sport* 11: 186–189.

Shostrom, E. 1968. *Man the manipulator.* New York: Bantam Books.

Simonton, D.K. 1990. Creativity in the later years: Optimistic prospects for achievement. *Gerontologist* 30(5) (October): 626–631.

Skinner, J.S., Tipton, G.M., and Vailas, A.C. 1982. Exercise, physical training and the aging process. In *Lectures in gerontology*, ed. A. Vijdik. New York: Academic Press, 407–439.

Slottje, D.J., Scully, G.W., Hirschberg, J.G., and Hayes, K.J. 1991. *Measuring quality of life across countries.* Boulder, CO: Westview Press.

Smith, D. 1973. *The geography of social well-being in the United States.* New York: McGraw-Hill.

Smith, S.T., Meyers, T., and Johnson, E. 1968. Simulation seeking as a function of duration and extent of sensory deprivation. *Proceedings of the 76th Annual Convention of the A.P.A.*: 625–662.

Sneider, M. 1975. The quality of life in large American cities: Objective and subjective social indicators. *Social Indicators Research* 1: 495–509.

Spirduso, W.W. 1986. Physical activity and the prevention of aging. In *Physical activity and well-being*, ed. V. Seefeldt. Reston, VA: Alliance of Health, Physical Education, Recreation and Dance.

Spirduso, W.W., and Clifford, P. 1978. Replication of age and physical activity effects for reaction and movement time. *Journal of Gerontology* 33: 26–30.

Sproull, N.L. 1988. *Handbook of research methods: A quide for practitioners and students in the social sciences.* Metuchen, NJ: Scarecrow Press.

Szalai, A., and Andrews, F.M. 1980. *The quality of life: Comparative studies.* London: Sage Publications.

Thomas, C., Kelman, H.R., Kennedy, G.J., Chul, A., and Chun-Yong, Y. 1992. Depres-

References

sive symptoms and mortality in elderly persons. *Journal of Gerontology* 47: 80–86.

Thurber, J. 1942. *My World and Welcome to It.* New York: Harcourt, Brace and World.

Toner, J., and Manuck, S.B. 1979. Health locus of control and health related information seeking at hypertension screening. *Journal of Social Science and Medicine* (Medical Psychology and Medical Sociology) 13A(6): 823–825.

A tribute to Marian Anderson. 1989. *Ebony* 45 (November): 182–186.

Vitality for life. 1993. Washington, DC: American Psychological Association.

Wagner, R.C., Fitts, P.M., and Noble, M.E. 1954. Preliminary investigations of speed and load as a dimension of psychomotor tasks. *USAF Pers. Train. Res. Cent. Rep.* (54-45).

Walker, B., and Mohr, T. 1985. The effects of ongoing self-directed questioning on silent comprehension. Paper presented at the annual meeting of the Reading Research Conference.

Waller, K., and Bates, R. 1992. Health locus and self-efficacy beliefs in a healthy elderly sample. *American Journal of Health Promotion* 6(4) (March-April): 302–309.

Wallston, B., Wallston, K., Kaplan, G., and Maides, S. 1976. Development and validation of the health locus of control scale. *Journal of Consulting and Clinical Psychology* 44(4): 580–585.

Wolfensberger, W. 1981. The extermination of handicapped people in World War II Germany. *Mental Retardation* (February).

Zinar, S. 1990. Fifth graders recall of propositional content and causal relationships from expository prose. *Journal of Reading Behavior* 22(2): 181–199.

Index

Aging. *See* Elderly people
Ailments, 160–162
Aristotle, 137
Aspirations, 163–165
Assessing achievement, 104, 176
Atrophy, 158–160
Attention deficit disorder, 153

Basic unit of analysis, 84
Bias, 14, 93
Boredom and confusion, 148–155

Caregiving, 10–11
Causality, 109–111
Cause-effect. *See* Causality
Centers for Disease Control and Prevention (CDC), 15, 35–48, 81–84
Confusion, 148–155
Control: cycle, 103–105, 145–146; of decision-making, 129–131; defined, 118; of intake, 125–126; loss of, 123–131; as a motive, 118–121; of movement, 127–128; personal implications of, 120; of production, 128–129; self, 119; theory, 120
Control cycle, 103–105; in research, 114–115, 177

Control motive, 118–122; societal implications of, 131–135

Data analysis, 111–114
Decision-making, 86–87; in research, 105–109
Defense mechanisms, 161
Dependency, 122–123
Depression, 153
Deriving treatments, 104, 172–175
Diagnosing causes, 104, 177–178
Disability, 10–11, 123–131, 160–163; compensation for, 162–163; indicators, 32–40. *See also* Rehabilitation
Discovery, 101–102, 114, 141–142
Drugs, 152–153, 174–175

Education, 68, 75. *See also* Instruction
Elderly people, 10, 118, 120, 122, 123, 125, 160, 162–163, 165
Enhancement, 7–8, 163–165
Euthanasia, 12
Exercise, 158–160
Expectancy. *See* Prediction

Failure. *See* Loss of control
Fatigue, 162–163
Feedback loop, 103

Freedom, 65–66
Fulfillment. *See* Happiness
Functional ability, 9, 31–53

Generalizing, 92–95
Geriatrics. *See* Elderly people
Gerontology. *See* Elderly people
Goals. *See* Projecting improvements
Government, 65, 71

Handicapped. *See* Disability
Happiness, 164–165
Health, 158–165
Health Related Quality of Life (HRQL), 49, 62–63, 81
Holistic approach to quality of life, 13–15, 84–85, 157–178; achievement of improvement, 169; areas of concern, 165–168; dynamic factors, 168–177; levels of concern, 168; structural factors, 157–168; treatment of, 169–170; using the control cycle, 170–177
Human motivation, 117–136; in school, 135; in the workplace, 132–134
Human superiority, 137
Hypothesis testing, 100–103

Impediments to predictive ability. *See* Boredom and confusion
Implementing treatments, 104, 175–176
Improvement, 158, 166–167. *See also* Control cycle
Incentives, 133. *See also* Human motivation
Innovation, 141–142, 145–146
Instincts, 9, 17
Instruction, 77
Instruments. *See* Observation instruments; Self-report instruments
Intelligence, 137–156. *See also* Predictive ability
Invention, 141, 145–146
IQ (Intelligence Quotient), 138, 147–148

Kelly, George, 140

Level of concern, 168
Life extension, 5
Longevity, 5–7

Malnutrition, 158
Man-machine systems, 143
Measurement, 97
Memory, 124, 129–130
Mental disorders. *See* Psychological problems
Mental health, 149, 152–153, 160–165
Mental retardation, 138–139
Motivation. *See* Human motivation

National Center for Health Statistics, 41–47
National Institute of Health (NIH), 2

Objectives. *See* Projecting improvements
Objectivity, 95–100
Observation instruments, 31–35; development of, 106
Optimum predictability, 150–151, 153–154

Personal quality of life indicators, 31–49
Prediction. *See* Predictive ability; Hypothesis testing
Prediction theory, 139–153
Predictive ability: and control, 139; defined, 139; impediments to, 148–155; improving, 146–147; and innovation, 141–142; and mental competency, 143; personal implications of, 140–142; and quality of life, 140–141; societal implications of, 142–145; and statistics, 145, 147; and work, 143–144
Preservation. *See* Self-preservation
Probability, 145, 158
Problem solving, 13–17, 103–109, 168–177
Projecting improvements, 104, 170–172
Protection, 69, 76
Provisions, 70, 77
Psychological problems, 149, 152–153. *See also* Mental health
Psychotherapy, 151–153
Purposeful behavior, 9

Quality of life: areas of concern, 65–80, 165–168; as a field of study, 61–90; indicators, 19–56, 65–80, 166–167; in-

Index

dicators summarized, 72, 78; of individuals, 70–80; instruments, 31–56; issues, 8–12; in societies, 65–70
Quality Adjusted Life Years (QUALYs), 33–34, 47–48

Recovery, 159–163
Recreation, 69, 76
Recuperation. *See* Fatigue
Rehabilitation, 4, 10, 14–15, 34, 49–57, 71, 118, 123–131, 150–153, 158–163, 172–175
Reincarnation, 5
Reliability, 98
Remote access, 68–69, 75–76
Research, 88–90, 101–115; causal-comparative, 107–108; descriptive, 108; evaluative, 106–107; experimental, 107; levels of, 108; policy, 105–106; predictive, 108; types of, 111, 172, 173–174, 176
Rest. *See* Fatigue

Satisfaction. *See* Happiness
Science, 80, 91–116; challenge to, 15–17
Scientific method, 101–102
Self-control, 119
Self-preservation, 5–7, 158; life extension, 5; procreation, 6
Self-report instruments, 35–44
Societal quality of life indicators, 20–31
Specialization, 14–15
Statistics. *See* Data analysis

Tests. *See* Observation instruments
Theory, prediction. *See* Prediction theory
Treatments, 80–81, 84, 166; development of, 107

Validity, 98–99

Work, 67–68, 74, 75, 142–145

Years of healthy life, 47

About the Author

MYLES I. FRIEDMAN is Gambrell Professor of Educational Psychology at the University of South Carolina. He has spent his professional life conducting research and building scientific theories in the fields of education and the social sciences. The emphasis of his work has been on human excellence, its nature, and attainment. Among his earlier publications are *Taking Control: Vitalizing Education* (Praeger, 1993), *The Psychology of Human Control: A General Theory of Purposeful Behavior* (with G.H. Lackey, Jr.) (Praeger, 1991), and *Rational Behavior* (1975).

ISBN 0-275-96028-5

90000>